International Conference on Power and Energy Engineering (ICPEE 2017)

IOP Conference Series: Earth and Environmental Science
Volume 114

Ottawa, Canada
13 – 15 September 2017

ISBN: 978-1-5108-5833-6
ISSN: 1755-1307

Printed from e-media with permission by:

Curran Associates, Inc.
57 Morehouse Lane
Red Hook, NY 12571

proceedings.com

Some format issues inherent in the e-media version may also appear in this print version.

Printed by Curran Associates, Inc. (2018)

For permission requests, please contact the Institute of Physics
at the address below.

Institute of Physics
Dirac House, Temple Back
Bristol BS1 6BE UK

Phone: 44 1 17 929 7481
Fax: 44 1 17 920 0979

techtracking@iop.org

Additional copies of this publication are available from:

Curran Associates, Inc.
57 Morehouse Lane
Red Hook, NY 12571 USA
Phone: 845-758-0400
Fax: 845-758-2633
Email: curran@proceedings.com
Web: www.proceedings.com

Table of contents

Volume 114

2017 International Conference on Power and Energy Engineering

13–15 September 2017, Ottawa, Canada

Accepted papers received: 22 January 2018
Published online: 7 February 2018

Preface

2017 International Conference on Power and Energy Engineering

2017 International Conference on Power and Energy Engineering

Peer review statement

Papers

Energy and Power Engineering

Power quality improvement by using STATCOM control scheme in wind energy generation interface to grid

Sheeraz Kirmani and Brijesh Kumar.....1

Application of computational fluid dynamics (CFD) simulation in a vertical axis wind turbine (VAWT) system

Jui-Hsiang Kao and Po-Yuan Tseng.....11

Utilization of busted CFL in developing cheap and efficient segmented compact LED bulbs

N S Andres and R T Ponce.....20

Heat radiation approach for harnessing heat of the cook stove to generate electricity for lighting system and charging of mobile phone

Rodrigo C. Muñoz Jr and Chad Deo G. Manansala.....28

Experimental Research on Optimizing Inlet Airflow of Wet Cooling Towers under Crosswind Conditions

You Liang Chen, Yong Feng Shi, Jian Gang Hao, Hao Chang and Feng Zhong Sun.....36

Application of porous medium for efficiency improvement of a concentrated solar air heating system

Boonrit Prasartkaew.....46

Crosswinds Effect on the Thermal Performance of Wet Cooling Towers Under Variable Operating Conditions

You Liang Chen, Yong Feng Shi, Jian Gang Hao, Hao Chang and Feng Zhong Sun.....54

Investigating the formation of acid mine drainage of Toledo pyrite concentrate using column cells

Diosa Marie Aguila.....59

Electronic Technology and Communication Engineering

Research on Precision Tracking on Fast Steering Mirror and Control Strategy

Lin Di, Wu Yi-ming and Zhu Fan.....64

A Multipath Mitigation Algorithm for vehicle with Smart Antenna

Jing Ji, Jiantong Zhang, Wei Chen and Deliang Su.....72

Power Allocation and Outage Probability Analysis for SDN-based Radio Access Networks

Yongxu Zhao, Yueyun Chen and Zhiyuan Mai.....79

Design of c-band telecontrol transmitter local oscillator for UAV data link

Hui Cao, Yu Qu and Zuxun Song.....87

Techniques for Computation of Frequency Limited H_∞ Norm

Shafiq Haider, Abdul Ghafoor, Muhammad Imran and Malik Fahad Mumtaz.....94

Research on Abnormal Detection Based on Improved Combination of K - means and SVDD

Xiaohong Hao and Xiaofeng Zhang.....103

Structural Engineering and Mechanical Manufacturing

An Evaluation Method of Equipment Reliability Configuration Management

Wei Wang, Weijia Feng, Wei Zhang and Yuan Li.....109

Research and analysis on response characteristics of bracket-line coupling system under wind load

Zhao Jiayu and Sun Qing.....115

Experimental study on the connection property of full-scale composite member

Cao Panpan and Sun Qing.....123

Slicer Method Comparison Using Open-source 3D Printer

M K A Mohd Ariffin, N A Sukindar, B.T. H T Baharudin, C N A Jaafar and M I S Ismail.....130

Design of external vibration absorber for vibration suppression of milling cutter in processing

Ya-hui Zhang, Nian-song Zhang and Ai-min Wang.....134

Designing Universitas Indonesia Molina EV Bus Dashboard Using ECQFD and TRIZ

Muhammad Faiq Pradhila, Amalia Suzianti and Prilly Putri Adinda.....140

PREFACE

It is our great pleasure to welcome you to 2017 International Conference on Power and Energy Engineering (ICPEE 2017) which will be held in Crowne Plaza Gatineau-Ottawa, Ottawa, Canada during September 13-15, 2017. ICPEE 2017 is dedicated to issues related to Power and Energy Engineering.

The major goal and feature of the conference is to bring academic scientists, engineers, industry researchers together to exchange and share their experiences and research results, and discuss the practical challenges encountered and the solutions adopted. Professors from Canada are invited to deliver keynote speeches regarding latest information in their respective expertise areas. It will be a golden opportunity for the students, researchers and engineers to interact with the experts and specialists to get their advice or consultation on technical matters, sales and marketing strategies.

This proceeding presents a selection from papers submitted to the conference from universities, research institutes and industries. All of the papers were subjected to peer-review by conference committee members and international reviewers. The papers selected depended on their quality and their relevancy to the conference. The volume tends to present to the readers the recent advances in the field of power and energy engineering and various related areas, such as alternative energy and the environment, power transmission and genneration, regulations, management and economics, communication and information systems engineering, computer applications and software development, etc..

We would like to thank all the authors who have contributed to this volume and also to the organizing committee, reviewers, speakers, chairpersons, sponsors and all the conference participants for their support to ICPEE 2017.

Prof. Gordon Huang
University of Regina, Canada
Sept. 21, 2017

IOP Conf. Series: Earth and Environmental Science **114** (2018) 011002 doi:10.1088/1755-1315/114/1/011002

Conference Committee

Conference Chair

Prof. Hossam A.Gabbar, University of Ontario Institute of Technology (UOIT), Canada

Program Chair

Prof. ZhongChao Tan, University of Waterloo, Canada

Prof. Gordon Huang, Faculty of Engineering and Applied Science, University of Regina, Canada

Dr. Xifeng Xu, Beijing University of Posts and Telecommunications, China

Steering Committees Chairs

Prof. Ming J Zuo,University of Alberta, Edmonton, Alberta, Canada

Prof. Eric James Strauss, Michigan State University, USA

Technical Committee

Prof. Ming J Zuo,University of Alberta, Edmonton, Alberta, Canada

Prof. Eric James Strauss, Michigan State University, USA

Prof. Yongtae Ahn, Gyeongnam National University of Science and Technology, Korea

Prof. Saad Abdel-Hamid El-Syaed, Mechanical power engineering-Zagazig university, Egypt

Prof. Ahmed A. Hossam-Eldin, Alexandria University, Alexandria, Egypt

Prof. RODRIGO C. MUÑOZ, BATAAN PENINSULA STATE UNIVERSITY, Philippines

Prof. Mohd Asri Mat Teridi, The National University of Malaysia

Assoc. Prof. Hasan AYDOGAN, Mechanical Engineering Department, Selcuklu, Konya, TURKEY

Assoc. Prof. Bal Raj Deshwal, Dept. of Chemistry, A. I. Jat H. M. College, Rohtak, India

Assoc. Prof. Sunil Kumar Singal, Alternate Hydro Energy Centre, Indian Institute of Technology, INDIA

Dr. Bandar Jubran Alqahtani, Duke University, USA

Dr. Hedayat Omidvar, Research & Technology Dept., National Iranian Gas Company, Iran

Dr. Jagdish Prasad Sharma, JK Lakshmipat University, Jaipur, Rajasthan, India

Dr. Hui Kwun Nam, University of Macau, China

Dr. Nugroho Agung Pambudi,Sebelas Maret University, Indonesia

Dr. Wei Peng, University of Regina, Canada

Dr. RAJA SEKHAR DONDAPATI, Lovely Professional University, India

Dr. Ipek F. Barut, Istanbul University, Turkey

Dr. Tsung-Mou Huang, Taiwan Power Company, Taiwan

IOP Conf. Series: Earth and Environmental Science **114** (2018) 011003 doi:10.1088/1755-1315/114/1/011003

Peer review statement

All papers published in this volume of *IOP Conference Series: Earth and Environmental Science* have been peer reviewed through processes administered by the proceedings Editors. Reviews were conducted by expert referees to the professional and scientific standards expected of a proceedings journal published by IOP Publishing.

IOP Conf. Series: Earth and Environmental Science **114** (2018) 012001 doi:10.1088/1755-1315/114/1/012001

Power quality improvement by using STATCOM control scheme in wind energy generation interface to grid

Sheeraz Kirmani[1] and Brijesh Kumar[2]

[1]Department of Electrical Engineering, Jamia Millia islamia, New Delhi, India. sheerazkirmani@gmail.com
[2]Department of Electrical Engineering, Jamia Millia islamia, New Delhi, India. bknsea@icloud.com

Abstract—"Electric Power Quality (EPQ) is a term that refers to maintaining the near sinusoidal waveform of power distribution bus voltages and currents at rated magnitude and frequency". Today customers are more aware of the seriousness that the power quality possesses, this prompt the utilities to assure good quality of power to their customer. The power quality is basically customer centric. Increased focus of utilities toward maintaining reliable power supply by employing power quality improvement tools has reduced the power outages and black out considerably. Good power quality is the characteristic of reliable power supply. Low power factor, harmonic pollution, load imbalance, fast voltage variations are some common parameters which are used to define the power quality. If the power quality issues are not checked i.e. the parameters that define power quality doesn't fall within the predefined standards than it will lead into high electricity bill, high running cost in industries, malfunctioning of equipments, challenges in connecting renewable. Capacitor banks, FACTS devices, harmonic filters, SVC's (static voltage compensators), STATCOM (Static-Compensator) are the solutions to achieve the power quality. The performance of Wind turbine generators is affected by poor quality power, at the same time these wind power generating plant affects the power quality negatively. This paper presents the STATCOM-BESS (battery energy storage system) system and studies its impact on the power quality in a system which consists of wind turbine generator, non linear load, hysteresis controller for controlling the operation of STATCOM and grid. The model is simulated in the MATLAB/Simulink. This scheme mitigates the power quality issues, improves voltage profile and also reduces harmonic distortion of the waveforms. BESS level out the imbalances caused in real power due to intermittent nature of wind power available due to varying wind speeds.

1. Introduction

Power is the very crucial input for the growth of any economy. Therefore, it is considered as a core industry as it facilitates development across various sectors, such as agriculture, manufacturing, railways, education, commercial etc. to expel economic development. To meet the high GDP rates the energy needs of the country are inevitable.

Renewable energy (RE) is characterised as intermittent and variable which presents various challenges in its grid integration for maintaining grid stability and security.

Intermittent/variable nature of RE source in an area of high penetration results in wide variation in quantum and direction of power flow on the inter-state high capacity transmission corridors. This requires placement of dynamic reactive compensation in the form of dynamic reactive compensation in the form of STATCOM/ SVC at strategic locations to provide dynamic support for smooth operation and maintaining grid security.

The integration of wind energy into existing electrical power system induces power quality problems like voltage regulation, stability, harmonic distortion, voltage sag/swell and poor power factor. The power quality is mainly customer-focused measure and is greatly affected by the operation of a distribution and transmission network. In this proposed scheme one of the Flexible AC transmission system (FACTS) device i.e., STATCOM is connected at point of common coupling (PCC) with a battery energy storage system (BESS) to mitigate power quality problems. Since, STATCOM connected to the grid provides reactive power support to wind generator as well as to loads. The BESS is integrated to sustain real power source under fluctuating wind or solar power.
In event of sudden load change or change in voltage profile during short circuit at point of common coupling STATCOM responds fast and stabilises the voltage and also helps to maintain power quality norms during such sudden and unexpected challenges.

2. Problems related to power quality

2.1. Power quality issues at grid side:
At the grid side the power quality is the responsibility of utility. Utility should make sure that the power matches the customer requirements and should not violate the limits that are specified for the parameters which define the power quality. From the customer point of view the voltage variations and high content of harmonics in the grid power are highly undesired as they affect the performance of the end equipments. For the IIP's who have planned the wind power project, the voltage profile of evacuating substation and nearby substations is of prime concern.
1) *Voltage variation:*
Intermittent nature of wind power causes several problems and one is variation of voltage of buses in the region of high RE penetration. Wind generators mostly employed induction generators and power electronic circuits which demand reactive power for operation. Voltage sag/swell is observed where ineffective methods of reactive power management are employed. If voltage rises beyond the controllable limit, forced tripping of lines carried out, cascaded tripping may destabilise a weak power system. Generally the power factor of evacuating substation is maintained near to unity preferably slightly lagging.
2) Voltage Transient:
Fault in the power system network, capacitor switching and HVDC systems are the primary cause of voltage transients. Voltage transients are responded well by STATCOM.

2.2. Power quality issues of WTG side
In wind energy generating system the power quality primarily concerned with the quality of current waveform which is being drawn or generated by the wind turbine. Poor power quality affects the performance of the loads connected to the grid.
1) *Reactive Power Consumption:*
Induction generators draw reactive power to produce its working flux while generate active power at the same time. As induction generators are most widely preferred in wind turbine generators, collectively a wind farm demand huge amount of reactive power.
As the wind speed is not constant, the use of electronic power conversion devices in wind turbine generators becomes inevitable to achieve a rotor speed for maximum extraction of energy from wind. The operation of power electronic devices also requires reactive power. To avoid voltage stability problem either STATCOM or capacitor arrangement is used to supply this demand of reactive power.
2) Current Harmonics Generation:
Capacitors are used as an essential part of the wind turbine generators for supplying reactive power demand. Capacitor switching may cause large voltage transient. The frequency and amplitude of such transient are enormous, particularly when back to back switching is involved, for instance capacitor bank switching. The over voltages may damage the insulation, Moreover, electronic equipments such as controllers are very sensitive to these transients, may produce incorrect commands. In addition, lightning strikes will cause an over voltage in the electrical system of wind turbine.

3. Topology for power quality improvement

The STATCOM based current control voltage source inverter injects the current into the grid in such a way that the source current are harmonic free and their phase-angle with respect to source voltage has a desired value. The injected current will cancel out the reactive part and harmonic part of the load and induction generator current, thus it improves the power factor and the power quality. To accomplish these goals, the grid voltages are sensed and are synchronised in generating the current command for the inverter. The proposed grid connected system is implemented for power quality improvement at point of common coupling (PCC), as shown in Figure 1. The grid connected system in Figure 1, consists of wind energy generation system and battery energy storage system with STATCOM.

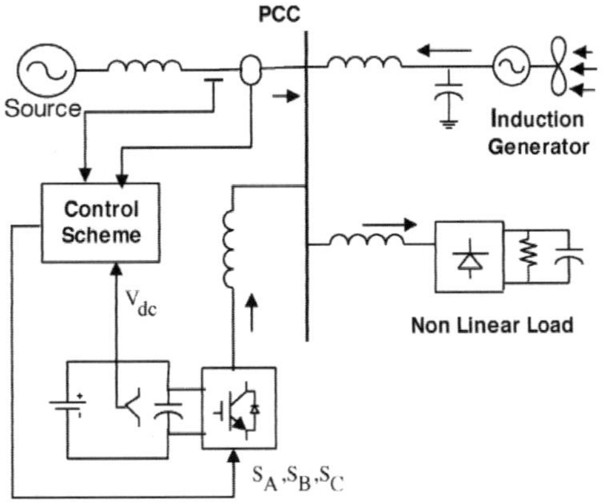

Figure 1. Diagrammatic representation of operating scheme

3.1. BESS-STATCOM

The battery energy storage system (BESS) is used as an energy storage element to support the wind farm during intermittencies it also support grid during any disturbance and loss of generation. The BESS will naturally maintain dc capacitor voltage constant and is best suited in STATCOM since it readily manages demand and supply of real power and also injects or absorbed reactive power to stabilise the grid system. It also controls the distribution and transmission system at a very fast rate. When power fluctuations occurs in the system, the BESS can be used to level the power fluctuations by charging and discharging operation. The BESS system is connected in parallel to the dc capacitor of STATCOM [10]–[14].

STATCOM comes from the family of FACTS devices. These are basically solid-state devices which are having the capability to respond to the reactive power demand. STATCOM have the edge over the SVC's as the former have constant current characteristics while in the SVC's the capacitive current drops linearly with the voltage. STATCOM can easily be interfaced with real power sources like the battery systems, fuel cells etc. STATCOM effectively control the system voltage and avoid voltage collapse.

STATCOM are solid state shunt connected devices. STATCOM's strategically placed in the power system to make the grid robust to the disturbances. STATCOM are finding applications in the renewable energy integration.

3.2. System operation

In the system under study STATCOM is interfaced with the BESS system. The STATCOM-BESS system is then connected to the PCC in the grid where non-linear loads and induction generator based wind turbine are also interfaced.Current control strategy is adopted to control the STATCOM -BESS system. The control strategy controls the output of STATCOM in such a manner so as to achieve power quality norms in the electrical grid. The STATCOM is intended here to support both reactive as well as real power demand of the other sub-systems.

4. Control scheme

IOP Conf. Series: Earth and Environmental Science **114** (2018) 012001 doi:10.1088/1755-1315/114/1/012001

4.1. Bang-bang current controller

The current control scheme is implemented using a bang- bang current controller. In this control scheme, the source current are detected by a current sensor and these are compared with the reference current to obtain the current error for the hysteresis based bang- bang controller.

Thus the ON/OFF switching signals for IGBT of STATCOM are derived from hysteresis controller. The switching function S_A for phase 'a' is expressed as.

$i_{sa} < (i_{sa}^{*} - HB) \rightarrow S_A = 0$

$i_{sa} > (i_{sa}^{*} - HB) \rightarrow S_A = 1$

where HB is a hysteresis current band, similarly the switching functionS_B ,S_C can be derived for phases "b" and "c"[14][15].

4.2. Gird,synchronisation

In three-phase balance system, the RMS voltage source amplitude is calculated at the sampling frequency from the source phase voltage (V_{sa}, V_{sb}, V_{sc}) and is expressed as sample template V_{sm}, sampled peak voltage, as in (1).

$$V_{sm} = \sqrt{\{\tfrac{2}{3}(V_{sa}^2 + V_{sb}^2 + V_{sc}^2)\}} \quad (1)$$

The in-phase unit vectors are obtained from AC source-phase voltage and the RMS value of unit vector u_{sa}, u_{sb}, u_{sc} as shown in (2).

$$u_{sa} = \frac{V_{sa}}{V_{sm}},\ u_{sb} = \frac{V_{sb}}{V_{sm}},\ u_{sc} = \frac{V_{sc}}{V_{sm}} \quad (2)$$

The in-phase generated reference currents are derived using in-phase unit voltage template as, in (3)

$$i_{sa} = I.u_{sa},\ i_{sb} = I.u_{sb},\ i_{sc} = I.u_{sc} \quad (3)$$

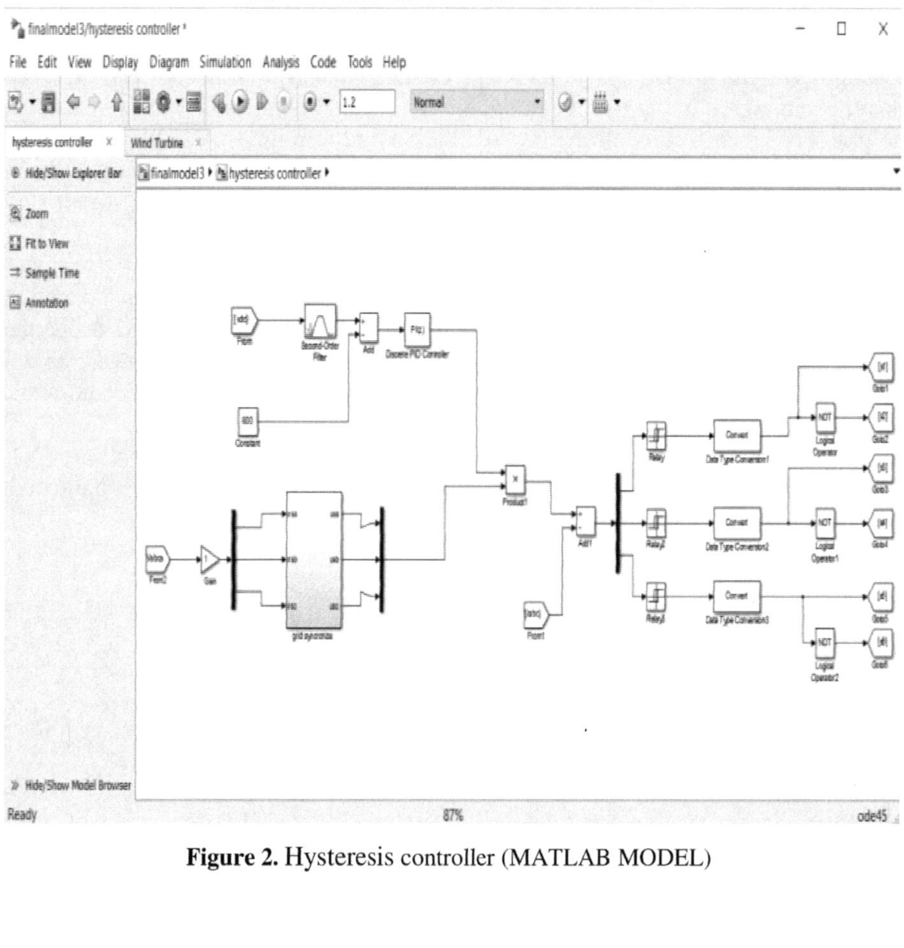

Figure 2. Hysteresis controller (MATLAB MODEL)

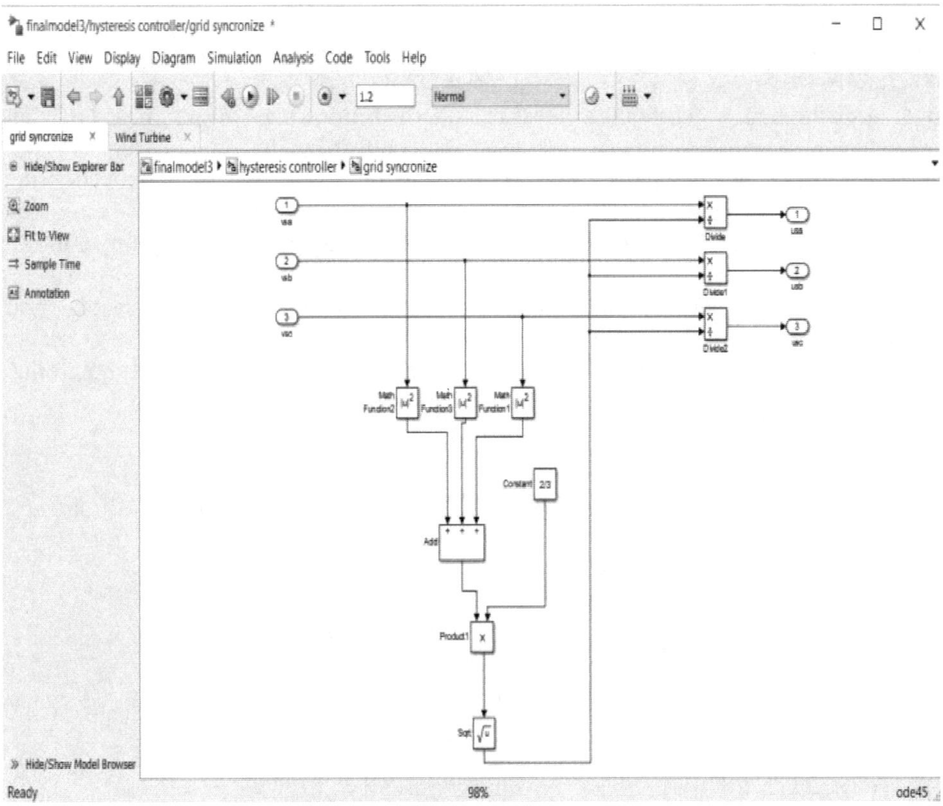

Figure 3. Unit Vector Block for Grid Synchronisation

Where *I* is proportional to the magnitude of filtered source voltage for respective phases. This ensures that the source current is controlled to be sinusoidal. The unit vectors implement the important function in the grid connection for the synchronisation for STATCOM. This method is simple, robust and favourable as compared with other methods [16].

5. System performance

The proposed control scheme is simulated using Simulink in power system block set. The system parameter for given system is given Table I.

Table 1

S.N.	Parameters	Ratings
1	Grid Voltage	3-phase ,415V,50 Hz
2	Induction Motor/Generator	3.35 kVA,415V, 50 Hz, P = 4, Speed = 1440 rpm, Rs = 0.01Ω, Rr=0.015Ω,Ls =0.06H,Lr=0.06H
3	Line Series Inductance	0.05mH
4	Inverter Parameters	DC Link Voltage = 800V, DC link Capacitance = 100 μF. Switching frequency = 2 kHz,
5	IGBT Rating	Collector Voltage =1200V, Forward Current =50A,Gate voltage =20V, Power dissipation = 310W
6	Load Parameter	Non-linear Load 25kW.

5.1. Voltage source current control—inverter operation

STATCOM in its basic structure is a voltage source inverter and is composed of IGBT's. IGBT based three phase inverter is connected to the grid at PCC through the interconnecting transformers. STATCOM inject three phase currents to level out the distortions due to non linear loads and wind turbine generating system in the grid power supply. The control signals with a hysteresis band of 0.08 for switching of IGBT's are simulated from the comparison made between reference currents and actual three phase source currents. Hysteresis band is generally kept low for better combating capabilities of STATCOM. The control signal of switching frequency within its operating band is shown in Figure 4.

The choice of the current band depends on the operating voltage and the interfacing transformer impedance. The compensated current for the nonlinear load and demanded reactive power is provided by the inverter. The real power transfer from the batteries is also supported by the controller of this inverter. The three phase inverter injected current are shown in Figure 5.

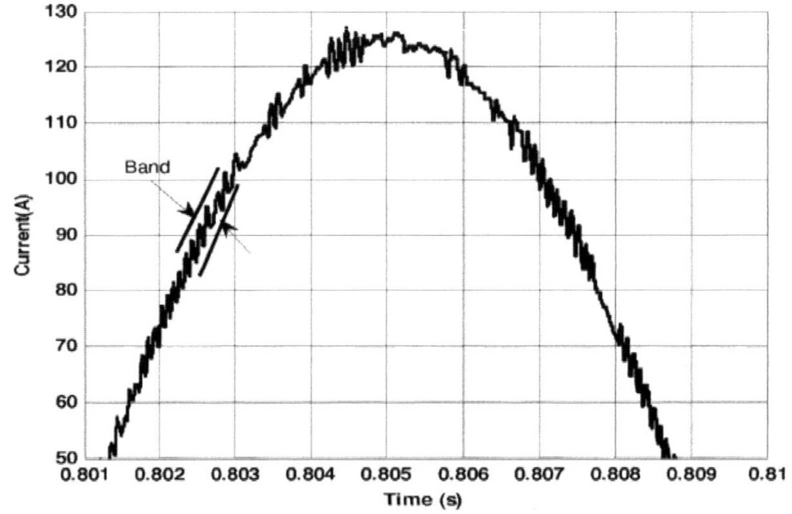

Figure 4. Switching signal and control hysteresis band.

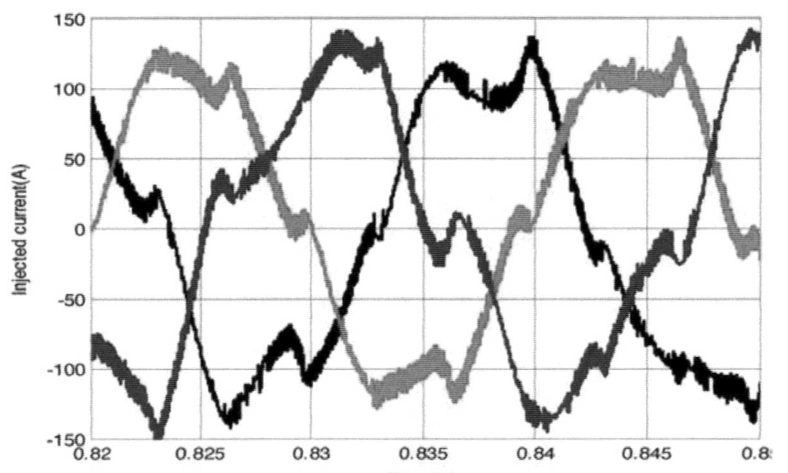

Figure 5. Three phase injected inverter Current.

5.2. STATCOM—performance under load variations

The system is modelled in MATLAB to investigate the performance of STATCOM based BESS system under dynamic loads and peak wind generation or fixed wind generation. Induction generator based wind turbine generator demand reactive power. Therefore when the STATCOM is switched on at time 0.7sec it start supplying this reactive power demand. Additionally it also injects current such that harmonics at the supply-side current reduces considerably. To investigate the performance under varying loads the provisions are made in the MATLAB program to increase the load at time 1sec. For increased load, STATCOM based system effectively compensate the increased demand of real and reactive power. The result of source current, load current are shown in Figure 6(a) and (b) respectively. While the result of injected current from STATCOM is shown in Figure 6(c) and the generated current from wind generator at PCC are depicted in Figure 6(d).

The DC link voltage regulates the source current in the grid system, so the DC link voltage is maintained constant across the capacitor as shown in Figure 7(a). The current through the DC link capacitor indicating the charging and discharging operations is shown in Figure 7(b).

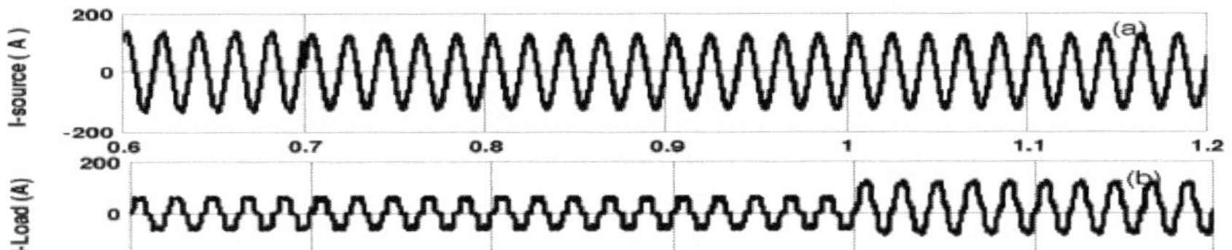

Figure 6. (a) Source Current. (b) Load Current. (c) Inverter Injected Current (d) Wind generator (Induction generator) current.

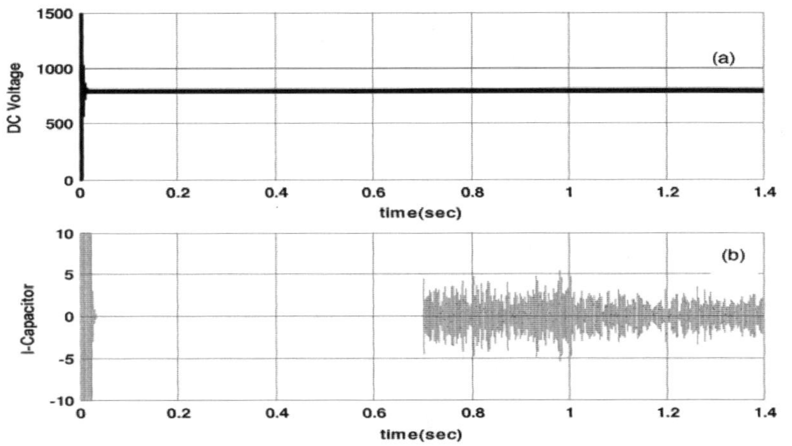

Figure 7. (a) DC link voltage. (b) Current through Capacitor.

5.3. Power quality improvement

Power quality is affected by the non linear loads as such loads inject harmonics in the grid. Wind turbine generators also have negative impact on power quality. Thus the purity good power quality on grid side and wind generator side is a challenge for the grid operators. STATCOM is observed to be satisfactorily compensating for the deteriorating power quality. The STATCOM output voltage and injected current are adjusted as per the power quality parameters sensed by the hysteresis controller from the grid. The high speed operation of controller ensures reliable power quality norms at the PCC. The STATCOM output voltage is shown in the figure 8. The source side voltage and current plot is shown in the figure 8, the plot clearly shows that power factor becomes near unity when the STATCOM is switched on. The harmonic analysis of the source side current before and after the beginning of STATCOM operation is carried out. When STATCOM is made on significant reduction of harmonics from initial value of 4.71 % is observed. The source current waveform when STATCOM is on is shown in the figure 10.

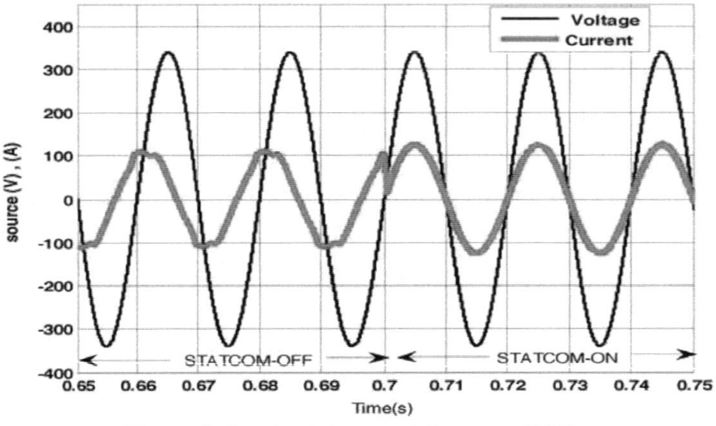

Figure 8. Supply Voltage and Current at PCC.

The system not only improves the power quality but also supports the loads with its battery system. It balances the supply and demand. This scheme also has the capabilities to support the loads in case of loss of generation in such case power from batteries can be supplied to the loads but for limited time.

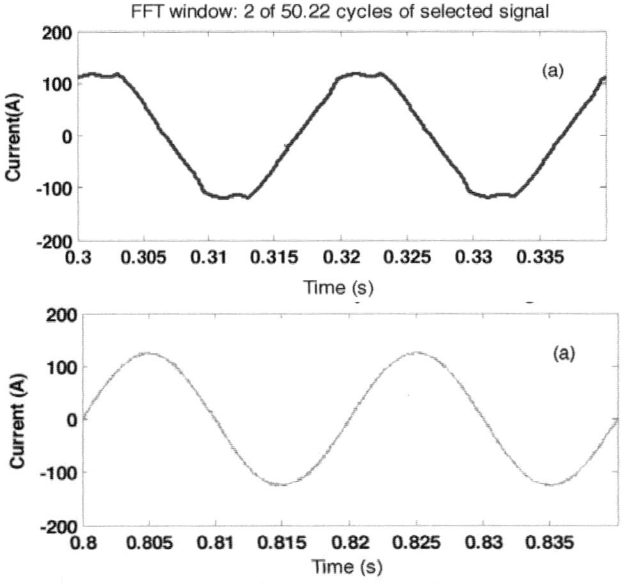

Figure 9. (a) Source Current (b) FFT of source current

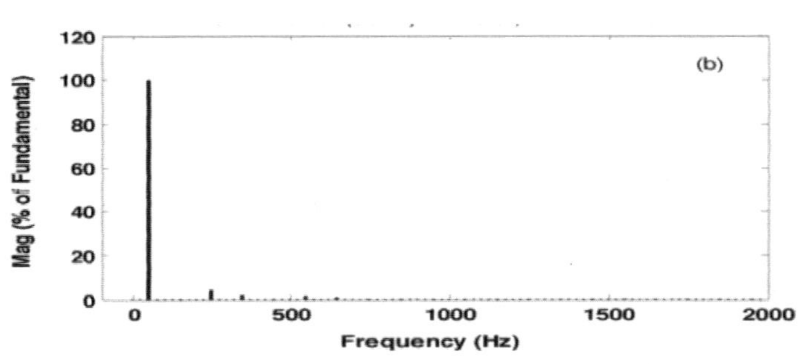

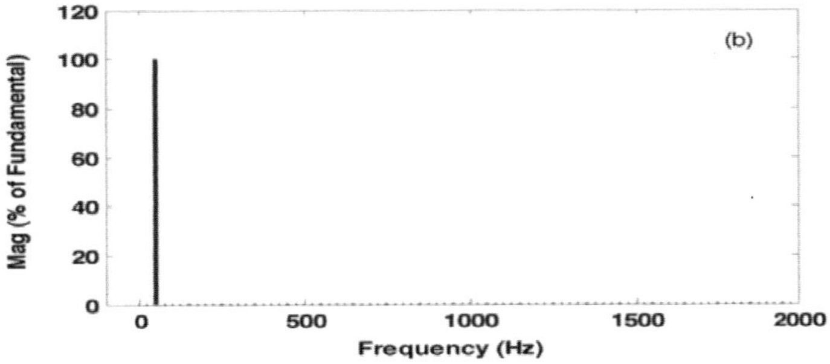

Figure 10. (a) Source Current (b) FFT of source current.

6. Conclusions

The paper presents STATCOM based battery energy storage system. The entire STACOM-BESS system is modelled in MATLAB/Simulink. When STATCOM is in operation significant improvement in power quality is observed. Issues like voltage sag\swell, harmonic distortion, power factor voltage profile are properly tackled. The current and voltage at source side are observed in phase after the STACOM is made on into the system. Therefore, a near unity power factor therefore can be maintained at the evacuating substation (PCC). In practice STATCOM assures power quality norms nearby substations of the wind generating farm. Therefore, forced tripping of connecting lines between pooling substation and evacuating substation can be avoided which improves the plant load factor and in tern revenues.

7. References

[1] K. S. Hook, Y. Liu, and S. Atcitty, "Mitigation of the wind generation integration related power quality issues by energy storage," *EPQU J.*, vol. XII, no. 2, 2006.

[2] J. Manel, "Power electronic system for grid integration of renewable energy source: A survey," *IEEE Trans. Ind. Electron.*, vol. 53, no. 4, pp. 1002–1014, 2006, Carrasco.

[3] C. Han, A. Q. Huang, M. Baran, S. Bhattacharya, and W. Litzenberger, "STATCOM impact study on the integration of a large wind farm into a weak loop power system," *IEEE Trans. Energy Conv.*, vol. 23, no. 1, pp. 226–232, Mar. 2008.

[4] T. Kinjo and T. Senjyu, "Output laveling of renewable energy by electric double layer capacitor applied for energy storage system," *IEEETrans. Energy Conv.*, vol. 21, no. 1, Mar. 2006.

[5] R. S. Bhatia, S. P. Jain, D. K. Jain, and B. Singh, "Battery energy storage system for power conditioning of renewable energy sources," in *Proc. Int. Conf. Power Electron Drives System*, Jan. 2006, vol. 1, pp. 501–506

[6] J. Zeng, C. Yu, Q. Qi, and Z. Yan, "A novel hysteresis current control for active power filter with constant frequency," *Elect. Power Syst. Res.*, vol. 68, pp. 75–82, 2004.

[7] J. Barros, M. de Apraiz, and R. I. Diego, "Measurement of Subharmonics In Power Voltages", Power Tech, IEEE Lausanne, Page(s): 1736 – 1740, 2007.

[8] Y. Lei, A. Mullane, G. Lightbody, and R. Yacamini, "Modeling of the wind turbine with a doubly fed induction generator for grid integration studies," IEEE Trans. Energy Conversion, vol. 21, no. 1, pp. 257–264, Mar. 2006.

[9] C. F. Lu, C. C. Liu, and C. J. Wu, "New dynamic models of lead-acid batteries," IEE Proc.-Gener. Trans. Distrib. , vol. 142, no. 4, pp. 429–435, Jul. 1995.

[10] Z. M. Salamah, M. A. Casacca, and W. A. Lynch, "A mathematical model for lead-acid batteries," IEEE Trans. Energy Conversion, vol. 7, no. 1, pp. 93–97, Mar. 1992.

[11] Z. Yang, C. Shen, L. Zhang, M. L. Crow, and S. Atcitty, "Integration of a STATCOM and battery energy storage," IEEE Trans. Power Syst., vol. 16, no. 2, pp. 254–260, May 2001.

[12] M. Black and G. Strbac, "Value of bulk energy storage for managing wind power fluctuations," IEEE Trans. Energy Conversion, vol. 22, no. 1, pp. 197–205, Mar. 2007.

[13] E. Spahic, G. Balzer, and A. D. Shakib, "The impact of the 'wind farm battery' unit on the

power system stability and control," in Proc. IEEE Power Tech., Lausanne, Jul. 2007, pp. 485–490.

[14] S. W. Mohod and M. V. Aware, "Grid power quality with variable speed wind energy conversion," in *Proc. IEEE Int. Conf. Power Electronic Drives and Energy System (PEDES)*, Delhi, Dec. 2006.

[15] S. W. Mohod and M. V. Aware, "Power quality issues & it's mitigation technique in wind energy conversion," in *Proc. of IEEE Int. Conf.Quality Power & Harmonic*, Wollongong, Australia, 2008.

[16] M. I. Milands, E. R. Cadavai, and F. B. Gonzalez, "Comparison of control strategies for shunt active power filters in three phase four wire system," *IEEE Trans. Power Electron.* , vol. 22, no. 1, pp. 229–236, Jan.2007.

[17] http://www.gwec.net/publications/country -reports/indian-wind-energy-outlook-2012/ [accessed on 01.02.16].

[18] http://www.gwec.net/wp-content/uploads/2012/11/Glob_cum_wind_power_cap_MW.jpg/ [accessed on 01.02.16].

[19] C. ZOU, B. WANG, and P. BAO, "Application of STATCOM in Wind Farm," Electric Drive, vol. 12, p. 010, 2008.

IOP Conf. Series: Earth and Environmental Science **114** (2018) 012002 doi:10.1088/1755-1315/114/1/012002

Application of computational fluid dynamics (CFD) simulation in a vertical axis wind turbine (VAWT) system

Jui-Hsiang Kao[1], and Po-Yuan Tseng[1]
[1]Department of Systems Engineering and Naval Architecture
National Taiwan Ocean University Keelung, Taiwan, R.O.C

Abstract. The objective of this paper is to describe the application of CFD (Computational fluid dynamics) technology in the matching of turbine blades and generator to increase the efficiency of a vertical axis wind turbine (VAWT). A VAWT is treated as the study case here. The SST (Shear-Stress Transport) k-ω turbulence model with SIMPLE algorithm method in transient state is applied to solve the T (torque)–N (r/min) curves of the turbine blades at different wind speed. The T–N curves of the generator at different CV (constant voltage) model are measured. Thus, the T–N curves of the turbine blades at different wind speed can be matched by the T–N curves of the generator at different CV model to find the optimal CV model. As the optimal CV mode is selected, the characteristics of the operating points, such as tip speed ratio, revolutions per minute, blade torque, and efficiency, can be identified. The results show that, if the two systems are matched well, the final output power at a high wind speed of 9–10 m/s will be increased by 15%.

1. Introduction

VAWTs are usually good choices for urban areas because the radiated noise is lower as compared with horizontal wind turbines. However, the generated mechanical power of VAWTs is unstable due to the fluctuated inflow around the rotor blades and the variation of inflow attack angles at different azimuth. The published researches generally centered on the hydrodynamic simulation of turbine blades and the logic setting of the controller; matching the two systems, the turbine blades and the generator, is rarely discussed. Accordingly, a matching process in which CFD technology is applied will be described to improve the efficiency of VAWTs.

CFD technology has been widely applied in turbine design. Hirsch (1990) [1] applied CFD in blade simulation. Bardina et al. (1997) [2] commented that the SST k-ω turbulence model derived by the Menter (1994) [3] was suitable for predicting the flows with separation under adverse pressure gradients. Since the 2000s, much attention has been paid to achieving higher efficiency for wind turbines. Eftichios et al. (2006) [4] applied the Perturb and Observe Algorithm to track the maximum power of the turbine system by the controller. Adam et al. (2007) [5] set Fuzzy Logic in the controller to obtain maximum power. Qing'an et al. (2015) [6] tested the effect of blade number on aerodynamic forces of a straight-bladed VAWT, and it was clarified that the power coefficient decreases with the increase of numbers of blades. Kosasih, and Hudin (2016) [7] improved the performance of the turbine system by detecting turbulence intensity. Yin et al. (2016) [8] analyzed the system design, basic dynamic characteristics, thermal losses and management aspects; hydro-viscous transmission was applied to enhance the overall efficiency. Natapol and Thananchai (2016) [9] presented a wind booster to enhance performance of the VAWT at both low and high wind speeds. In Jian et al. (2016) [10], an airfoil family was specified by a coupled approach. This airfoil family achieved higher efficiency than traditional NACA 0015 by 15.5%. Besides, it was verified that the thickness-chord ratio had the biggest effect on the blade efficiency. Abdolrahim et al. (2017) [11] simulated a VAWT by 2D unsteady Reynolds-averaged Navier-Stokes. It was verified the 3D tip effects are negligible as the aspect ratio of the turbine is high. Ghulam, and Uzma (2017) [12] applied magnetic levitation method to the axial flux permanent magnet generator of a VAWT for increasing the efficiency. Under the inspiration of migrating geese flying in a V or I formation to save energy, Baoshou et al. (2017) [13] presented a novel wake energy reuse method to optimize the Savonius-type VAWT. Qing'an et al. (2017) [14] investigated the effect of rotor aspect ratio and solidity on a straight-bladed VAWT by 3D Panel Method. It was concluded the peak of power coefficient increases with the increase of the ratio of the blade span length and diameter at the fixed solidity.

In this paper, the efficiency of the VAWT is improved by matching the turbine blades and the generator. The performance of the blades is predicted by CFD technology first. Then the predicted results are matched with the selected generator by the T-N curves to determine the optimal operating point at different wind speed. Both matched and unmatched results will be compared to prove the improvement in the system efficiency.

2. Mathematical model

The aerodynamics of the present study case is investigated by using the commercial CFD software, ANSYS FLUENT. The SST (Shear-Stress Transport) k-ω turbulence model of the second-order upwind scheme with the SIMPLE algorithm method is applied to solve the Reynolds-averaged Navier Stokes (RANS) equations. The time averaging approach for a time dependent and irregular motion splits all the time varying terms into mean and fluctuating values. ANSYS Fluent solves the RANS equations by finite volume discretization. The SST k-ω model introduced by Menter (1992) [15] is applied in this paper. To measure adverse pressure gradients, aerofoils and transonic shock waves, the SST k-ω model has been strongly recommended by Menter (1992) [16]. The SIMPLE algorithm derived by Patankar & Spalding (1972) [17] was used to solve the incompressible flow. The main idea of this scheme is to update the pressure and velocities by corrected terms.

3. The governing equations

Yuwei et al. (2012) [18] mentioned that the flow in wind turbines is still essentially incompressible with Mach numbers based on a blade tip speed rarely exceeding 0.25. The Mach numbers based on the blade tip speed of in this present study did not exceed 0.25; thus, it was reasonable to consider the flow as incompressible in this paper. With the assumptions of incompressible and turbulent-steady flow, mass and momentum conservation equations can be written respectively as below:

$$\frac{\partial \overline{u}_j}{\partial x_j} = 0 \tag{1}$$

$$\rho\frac{\partial \overline{u}_i}{\partial t} + \rho\overline{u}_j\frac{\partial(\overline{u}_i)}{\partial x_j} = -\frac{\partial \overline{P}}{\partial x_i} + \frac{\partial}{\partial x_j}\left(\mu\frac{\partial(\overline{u}_i)}{\partial x_j}\right) + \frac{\partial R_{ij}}{\partial x_j} \tag{2}$$

$\overline{u}_j$ is the average velocity along the x_j direction, ρ the fluid density, $\overline{P}$ average pressure, Rij the Reyonlds stress.

$$R_{ij} = -\rho\overline{u_i^{'}u_j^{'}} \tag{3}$$

3.1 The applied turbulent model

The two equations of SST k-ω model are as followings.

Turbulence Kinetic:

$$\frac{\partial(\rho k)}{\partial t} + \frac{d(\rho k u_i)}{dx_i} = \frac{\partial}{dx_j}\left(\Gamma_k\frac{dk}{dx_j}\right) + \overline{G_k} - Y_k + S_k \tag{4}$$

Specific Dissipation Rate:

$$\frac{\partial(\rho\omega)}{\partial t}+\frac{d(\rho\omega u_i)}{dx_i}=\frac{\partial}{dx_j}\left(\Gamma_\omega\frac{d\omega}{dx_j}\right)+G_\omega-Y_\omega+D_\omega+S_\omega \tag{5}$$

Here $\overline{G_k}$ is the turbulence kinetic energy caused by the mean velocity gradients, G_ω the generation of ω, Γ_k and Γ_ω the effective diffusivity of k and ω, Y_k and Y_ω the dissipation of k and ω due to turbulence, S_k and S_ω source terms, and D_ω the cross-diffusion term.

4. SIMPLE algorithm method

The SIMPLE algorithm is often used to solve the incompressible flow. The main ideal of this scheme is to update the pressure and velocities by corrected terms. In this scheme, the momentum equation is replaced by following discrete form.

$$a_n u_n' = \sum_{nb} a_{nb}u_{nb}' + A_n\left(P_P' - P_N'\right) \tag{6}$$

a_{nb} is the coefficient of the convective and diffusive terms summed by the nearby grid points. $A_n\left(P_P' - P_N'\right)$ presents the pressure force on the control volume. The corrections in velocity from the nearby grid points are assumed to be ignorable.

$$\sum_{nb} a_{nb}u_{nb}' = 0 \tag{7}$$

The corrected terms in velocities can be reformed.

$$\begin{aligned} u_n' &= d_n\left[P_P' - P_N'\right] = \left(\frac{A_n}{a_n}\right)\left[P_P' - P_N'\right] \\ v_e' &= d_e\left[P_P' - P_E'\right] = \left(\frac{A_e}{a_e}\right)\left[P_P' - P_E'\right] \\ w_u' &= d_u\left[P_P' - P_U'\right] = \left(\frac{A_u}{a_u}\right)\left[P_P' - P_U'\right] \end{aligned} \tag{8}$$

The corrected term in pressure can be calculated by following.

$$aP' = \sum_{nb} a_{nb}P_{nb}' \tag{9}$$

Thus, the updated pressure and velocities are as below.

$$\begin{aligned} P &= P^* + P' \\ u &= u^* + d_n\left(\Delta P'\right) \\ v &= v^* + d_e\left(\Delta P'\right) \\ w &= w^* + d_u\left(\Delta P'\right) \end{aligned} \tag{10}$$

5. Meshing arrangement and boundary conditions

A straight-bladed VAWT with 3 blades shown in Figure 1 is discussed in this paper. The rotor diameter (D) is 2.6 M, and blade length (H) is 2.6 M. The matching problem is emphasized here. Thus, the 3D tip effects are negligible, and 2D CFD simulation is carried out in this paper.

According to Steffen et al. (2001) [19], in order to correctly analyze the relative velocity of a rotating turbine, two computing domains are necessary; one encloses the calculated structures, and the other one includes the whole computing domain. Thus, the inner and outer domains are both developed in the present computation. Please refer to Figure 2 for the computing domains and the arrangement.

The applied meshes including 130000 mixed cells are shown in Figure 3. Figure 3 shows the meshes near the blade surface intensive. The mesh size increases from the blade toward the downstream and upstream directions.

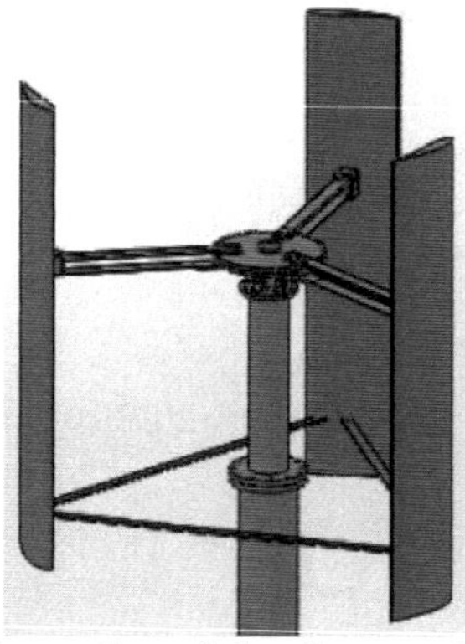

Figure 1. The outline of the discussed wind turbine

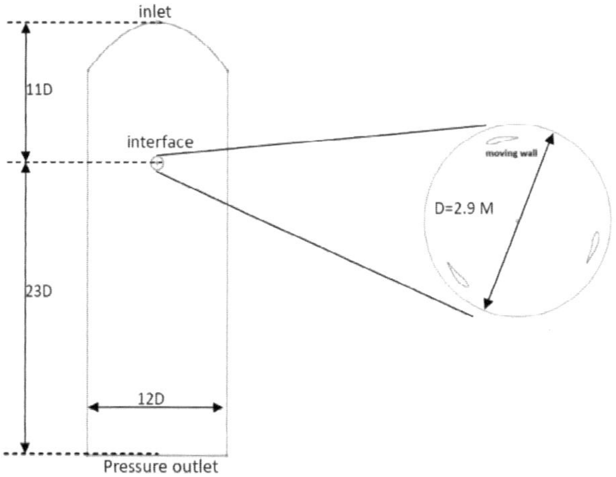

Figure 2. The computing domains and applied boundary conditions

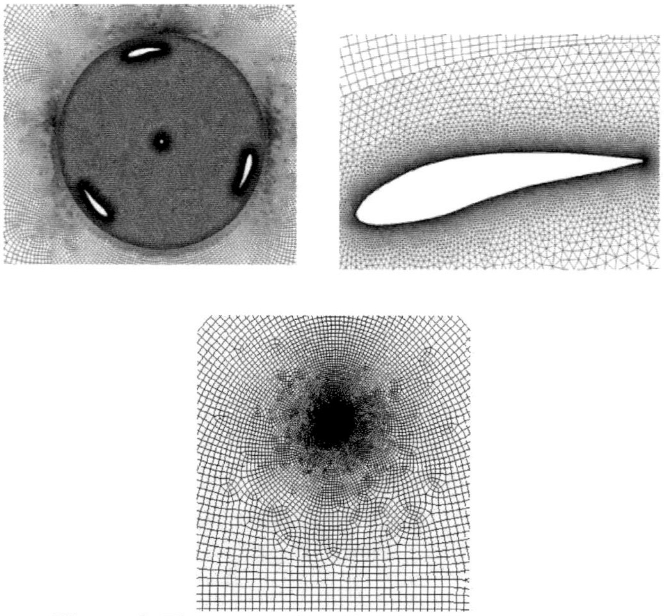

Figure 3. The arranged meshes for the present case

The boundary conditions for the present simulation are indicated in Figure 2. On the outer-domain inlet, a steady uniform velocity, V, is imposed and the outer-domain outlet is set as a pressure outlet. The outer-domain walls are considered as stationary with no slip. Rotational moving and no slip walls are applied in the inner domain.

6. Results and discussion

The selected generator, to be matched with the turbine blades, is a direct-drive type with a 3-phase permanent NdFeB magnet. The measured T-N curves of this generator with every CV model are plotted in Figure 4 and the corresponding efficiencies are shown in Figure 5. According to Figure 5, for any CV mode the maximal efficiency is larger than 80%; the slope varies rapidly before reaching the maximal efficiency point. For attaining high efficiency, the location of the operating point should not be in the steep-slope region.

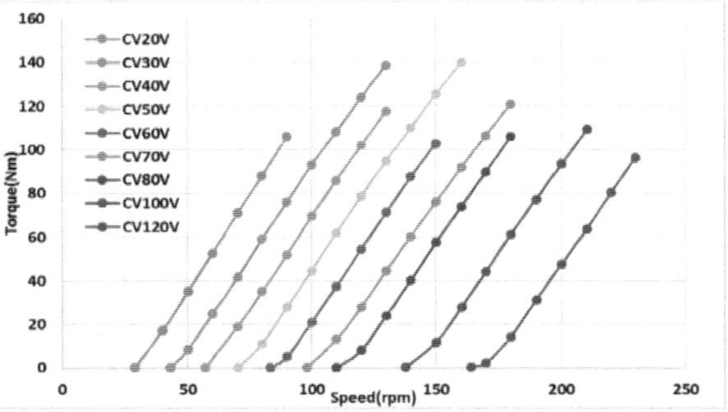

Figure 4. The T-N curves of this generator measured by different CV mode

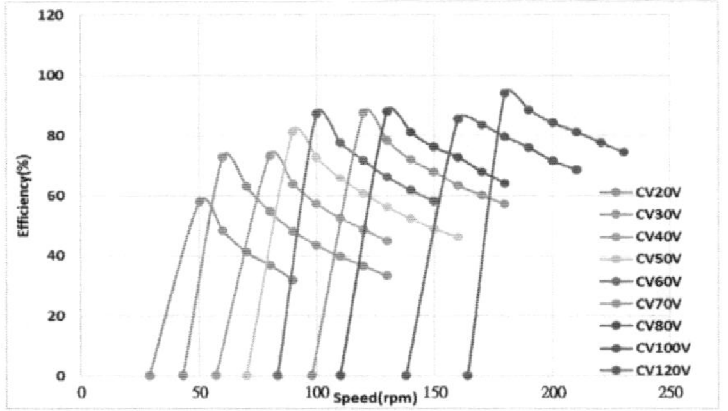

Figure 5. The efficiency of this generator at different CV model

6.1 The matched case

In this paper, the considered wind speeds are 5m/s, 6m/s, 7m/s, 8m/s, 9m/s and 10m/s. The T (torque)-N (RPM) curves of the turbine blades, according to the CFD simulation, are plotted in Figures 6 & 7. It is also evident in Figures 6 & 7 that when the inflow attack angle exceeds the stall angle, the blade torque will be reduced. The C_p-TSR curve is shown in Figure 8. C_p is the blade efficiency. TSR, tip speed ratio, is the ratio of the tip speed to the wind speed.

$$C_p = \frac{2 \cdot \pi \cdot N \cdot T}{\frac{1}{2} \cdot \rho \cdot V^3 \cdot (D \cdot H)} \tag{11}$$

$$TSR = \frac{2 \cdot \pi \cdot N \cdot R}{V} \tag{12}$$

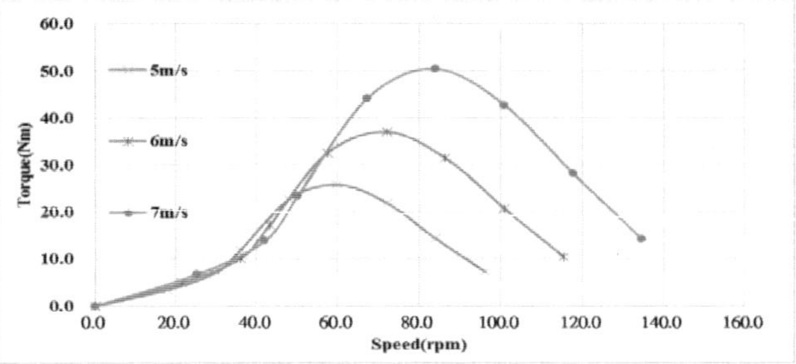

Figure 6. The T (torque)-N (RPM) curve of the blades at different wind speed

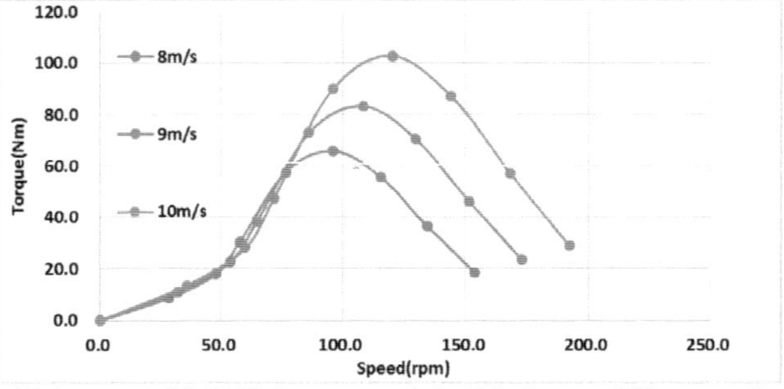

Figure 7. The T (torque)-N (RPM) curve of the blades at different wind speed

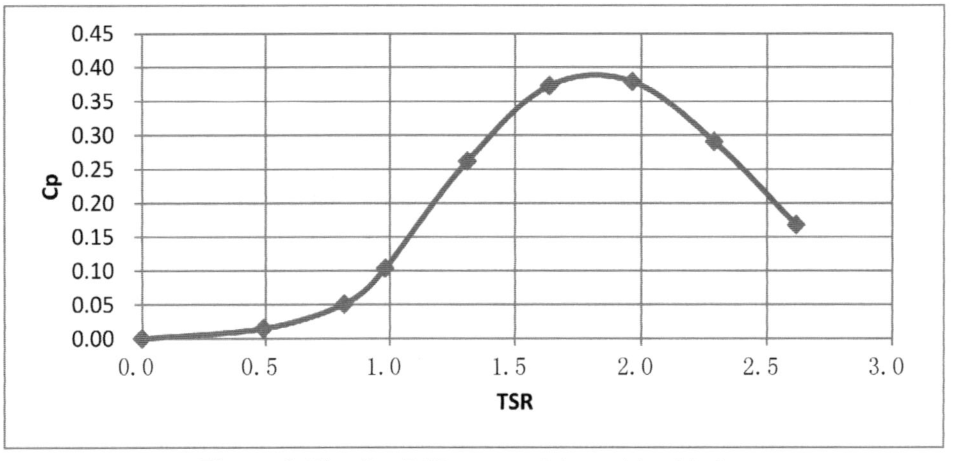

Figure 8. The Cp- TSR curve of the turbine blades

Figure 8 shows that when the TSR of the operating point for this wind turbine is located at 1.6~2.0, the blades will operate more efficiently

For a wind turbine system, the torque generated by the blades is used to overcome the torque of the generator. In this way, the mechanical energy can be transferred to electronic energy by the generator. Thus the data shown in Figures 6 & 7 should be matched with those shown in Figure 4. Figure 9 is obtained by combining Figures 6 and 4, and Figure 10 is obtained by combining Figures 7 and 4. From Figures 9 & 10, the operating point (i.e. the matched point) at every wind speed can be determined. At the optimal operating point, both the efficiencies of turbine blades and the generator should be kept high. The efficiency of the turbine blades varies with the dimensionless factor, as shown in Equation (12), TSR. As shown in Figure 8, when the TSR of the operating point for this wind turbine is located at 1.6~2.0, the blades will become more efficient. Besides, it should be noted that the operating speed, RPM, cannot be in the steep-slope region of the selected CV mode, as shown in Figure 3. In this steep-slope region, the efficiency of the generator will obviously be reduced. Thus, the criteria for the choice are operating points are the TSR-C_p curve as shown in Figure 8, the efficiency curves of the generator at every CV mode as shown in Figure 2, the determined operating speed, RPM, and the selected CV mode of the generator. It is expected that the TSR value calculated by the determined RPM will be located in the high-efficiency region of Figure 8, and not in the steep-slope region of the selected CV mode in Figure 3.

Table 1 shows the associated information of each matched operating point for every wind speed. Take 5 m/s wind speed for example. A CV 40V mode of generator is selected as the electronic control model. According to Figure 9, the matched point of 5m/s corresponds to 72 RPM and 21.7 Nm. From Figure 3 we can see that the generator efficiency, eff-g, of CV 40V at 72RPM is 79%. The electronic efficiency, eff-e, is usually 90~93%, and is assumed to be 90% here. The blade efficiency is about 37.9% as calculated by Equation (11). P_{-blade} (i.e. the mechanical power), 163.6W, and P_{-out} (i.e. the final electronic power), 116.3W, are computed by the following:

$$P_{-blade} = 2 \cdot \pi \cdot N \cdot T \tag{13}$$

$$P_{-out} = P_{-blade} \cdot eff_{-g} \cdot eff_{-e} \tag{14}$$

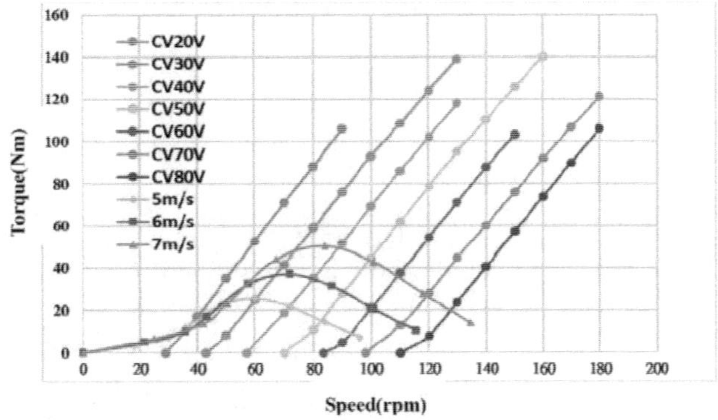

Figure 9. The matching results of turbine blades and generator

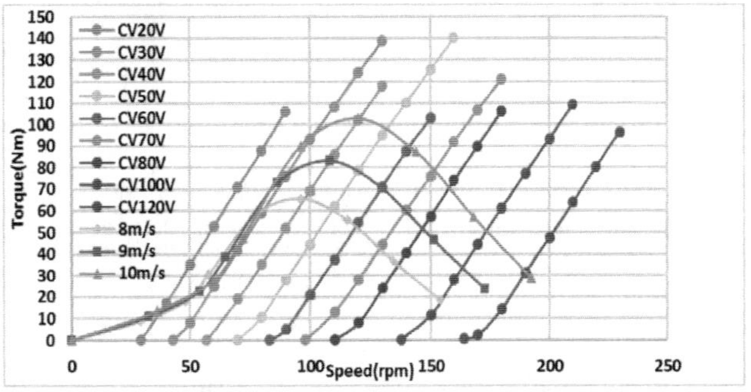

Figure 10. The matching results of turbine blades and generator

Table 1. The matched operating points at every wind speed

V(m/s)	5	6	7	8	9	10
CV	40	40	40	40	50	50
RPM	72	80	88.5	98	120	132
Torque (Nm)	21.7	33.9	48.3	64.8	76.3	95.0
eff-g	0.79	0.74	0.66	0.58	0.61	0.55
eff-e	0.90	0.9	0.9	0.9	0.9	0.9
Cp	0.379	0.376	0.374	0.374	0.376	0.376
P-blade	163.6	284.0	447.6	665.0	958.8	1313.2
P-out	116.3	189.1	265.9	347.1	526.4	650.0
TSR	1.96	1.81	1.72	1.67	1.81	1.80

6.2 The unmatched case

When the two systems: blade and generator, are unmatched, the generator will operate at the same CV mode for all wind speeds. In the case where these two systems are not matched, an overall CV mode should be selected first. Here the 40V mode is chosen in the unmatched case because the T-N curves of the turbine blades at every wind speed can form an intersection point with that of the 40V mode, as shown in Figure 11. At the intersection point, the torque needed by the generator is equal to that generated by the turbine blades.

The final outputted powers are listed in Table 2. By comparing the P-out listed in Tables 1 and 2, it can be seen that the final outputted power of the unmatched case at high wind speeds, 9~10 m/s, will be decreased by 15%.

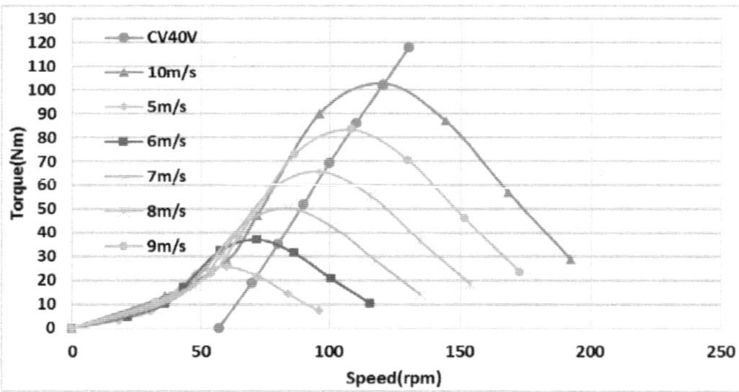

Figure 11. The selection of the CV mode for the unmatched case

Table 2. The unmatched operating points at every wind speed

V(m/s)	5	6	7	8	9	10
CV	40	40	40	40	40	40
RPM	72	80	88.5	98	108	120
Torque (Nm)	21.7	33.9	48.3	64.8	83.2	102.7
eff-g	0.79	0.74	0.66	0.58	0.53	0.49
eff-e	0.90	0.90	0.90	0.90	0.90	0.90
Cp	0.379	0.376	0.374	0.374	0.371	0.371
P-blade	62.0	106.8	167.4	248.7	941.0	1290.6
P-out	44.1	71.1	99.4	129.8	448.8	569.1
TSR	1.96	1.814222	1.720271	1.666817	1.63	1.63

7. Conclusions

The main purpose of this paper was to develop a matching process for vertical axis wind turbines to improve the efficiency. Several points can be concluded, as follows:

1. The basis for choosing the operating points are: the TSR-Cp curve of the turbine blades, the efficiency curves of the generator at every CV mode, the determined operating speed, RPM, and the selected CV mode of the generator. It is suggested that the determined RPM be located in the high-efficiency region of the TSR-Cp curve of the turbine blades, not in the steep-slope region of selected CV mode.
2. By comparing the final outputted power of the matched condition with these of the unmatched condition, it is known that the final outputted power of the matched condition at high wind speed, 9~10m/s, will be increased by 15%.
3. The proposed method of achieving an optimal balance of turbine and generator is still in the developmental stage. Over-speed protection and noise abatement during the matching process are suggested for future study. The expectation should be that the matched rotating speed is under control as the wind turbine experiences gusts, and does not cause unacceptable noise.

8. References

[1] Hirsch Charles. (1990) Numerical Computation of Internal and External Flows. Wiley.
[2] Bardina, J. E., Huang, P. G., & Coakley, T. J. (1997) Turbulence Modeling Validation, Testing, and Development. NASA Technical Memorandum No. 110446.
[3] Menter, F. R. (1994) Two-Equation Eddy-Viscosity Turbulence Models for Engineering Applications. AIAA J., Vol. 32, No. 8: 1598–1605.
[4] Eftichios Koutroulis, & Kostas kalaitzakis (2006) Design of a Maximum Power Tracking System for Wind-Energy-Conversion Applications. IEEE Transactions on Industrial Electronics, Vol. 53, No. 2: 486-494.
[5] Adam Mirecki, Xavier Roboam, & Frederic Richardiau (2007) Architecture Complexity and Energy Efficiency of Small Wind Turbines. IEEE Transactions on Industria l Electronics, Vol. 54, No. 1.
[6] Qing'an Li, Takao Maeda, Yasunari Kamada, Junsuke Murata, Kazuma Furukawa, & Masayuki Yamamoto (2015) Effect of number of blades on aerodynamic forces on a straight-bladed Vertical Axis Wind Turbine. Energy, Vol. 90, Part 1: 784–795.
[7] Kosasih, B., & Hudin, H. S. (2016) Influence of Inflow Turbulence Intensity on The Performance of Bare And Diffuser-Augmented Micro Wind Turbine Model. Renewable Energy, Vol. 87, Part 1: 154–167.
[8] Yin, X. X., Lin, Y. G., Li, W., & Gu, H. G. (2016) Hydro-viscous Transmission Based Maximum Power Extraction Control For Continuously Variable Speed Wind Turbine With Enhanced Efficiency. Renewable Energy, Vol. 87, Part 1: 646–655.
[9] Natapol Korprasertsak, & Thananchai Leephakpreeda. (2016) Analysis and optimal design of wind boosters for Vertical Axis Wind Turbines at low wind speed. Journal of Wind Engineering and Industrial Aerodynamics, Vol. 159, pp. 9–18.
[10] Jian Chen, Liu Chen, Hongtao Xu, Hongxing Yang, Changwen Ye, & Di Liu (2016). Performance Improvement of a Vertical Axis Wind Turbine by Comprehensive Assessment of an Airfoil Family. Energy, Vol. 114: 318–331.
[11] Abdolrahim Rezaeiha, Ivo Kalkman, & Bert Blocken (2017) CFD Simulation of a Vertical Axis Wind Turbine Operating at a Moderate Tip Speed Ratio: Guidelines For Minimum Domain Size And Azimuthal Increment. Renewable Energy, Vol. 107: 373–385.
[12] Ghulam Ahmad, & Uzma Amin (2017) Design, Construction And Study of Small Scale Vertical Axis Wind Turbine Based on a Magnetically Levitated Axial Flux Permanent Magnet Generator. Renewable Energy, Vol. 101: 286–292.
[13] Baoshou Zhang, Baowei Song, Zhaoyong Mao, & Wenlong Tian (2017) A Novel Wake Energy Reuse Method to Optimize The Layout For Savonius-Type Vertical Axis Wind Turbines. Energy, Vol. 121: 341–355.
[14] Qing'an Li, Takao Maeda, Yasunari Kamada, Kento Shimizu, Tatsuhiko Ogasawara, Alisa Nakai, & Takuji Kasuya (2017) Effect of Rotor Aspect Ratio And Solidity on a Straight-Bladed Vertical Axis Wind Turbine in Three-Dimensional Analysis by The Panel Method. Energy, Vol. 121: 1–9.
[15] Menter F. R. (1992) Performance of Popular Turbulence Model For Attached And Separated Adverse Pressure Gradient Flows. AIAA journal, Vol. 30, No. 8: 2066-2072.
[16] Menter F. R. (1992) Improved Two-Equation K-Omega Turbulence Models For Aerodynamic Flows. NASA TM 103975.
[17] Patankar S. V., & Spalding, D.B. (1972) A Calculation Procedure For Heat, Mass And Momentum Transfer in Three-Dimensional Parabolic Flows. Int. J. of Heat and Mass Transfer, Vol. 15, issue 10, October 1972: 1787-1806.
[18] Yuwei Li, Kwang-Jun Paik, Tao Xing, & Pablo M. Carrica (2012) Dynamic Overset CFD Simulations of Wind Turbine Aerodynamics. Renewable Energy, Vol. 37, issue 1: 285-298.
[19] Steffen Wußow, Lars Sitzki, & Thomas Hahm (2001) 3D-Simulation of The Wake of a Wind Turbine. DEWI Magazine Nr. 18.

IOP Conf. Series: Earth and Environmental Science **114** (2018) 012003 doi:10.1088/1755-1315/114/1/012003

Utilization of busted CFL in developing cheap and efficient segmented compact LED bulbs

N S Andres[1] and R T Ponce[1]

[1]Electrical Engineering Department, Bataan Peninsula State University, 2100, Balanga City Bataan, Philippines

Abstract. Today's generation will not survive a day without the help of lighting. In fact, someone's productivity, particularly at night, depends on the presence of a good lighting and it seems that it is a daily necessity. Lighting takes a large part on the consumption of household electrical energy particularly in the Philippines. There are different type of lighting bulbs used at home can affect the overall lighting consumption. Nowadays, most commonly and widely used bulb in the household is the Compact Fluorescent Light (CFL). However, the main problem of CFL is the mercury they contain. In addition to this is the harmful effect of mercury such as Emission of UV Radiation. In response to the said problem, this project study gives solution to the problem of the society concerning environment, health and safety as well energy conservation, by developing a segmented compact light-emitting diode (SCLED) bulb from busted CFL that are efficient, economical, and does not contain toxic chemicals.

1. Introduction

Based on the results of the 2011 Household Energy Consumption Survey (HECS), electricity remains as the most common source of energy used by households particularly in the Philippines. About 87 percent of 21.0 million households used electricity from March to August 2011. As of 2016, the demand for electricity in the residential is leading among the different sectors. It is often grouped depending on its uses that are distributed for heating and cooling, lighting, operating appliances etc [1]

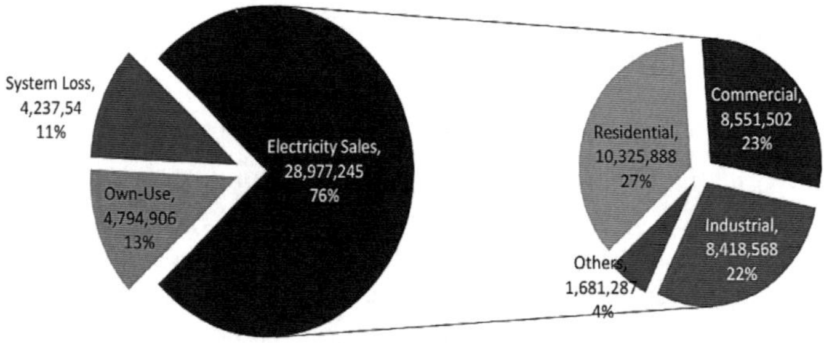

Figure 1. January- june 2016 philippine power demand

Now that the world is in the age where lighting seems to be a daily necessity, typical homes as shown in figure 1, consume nearly 27 percent of the energy used today: making lighting as the major source of electricity consumption. Lighting plays a large part on the consumption of household electrical energy; it consumes about 18% of the total generated electricity of total energy consumption in residential sectors as shown in figure 2 [2].

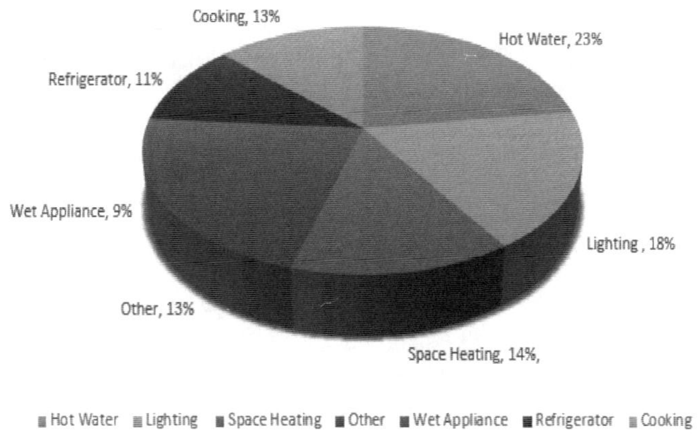

Figure 2. Estimated electricity use in residential sector

One of the main factors in lighting energy consumption is the light bulb. The use of correct and appropriate type of light bulb improves the efficiency of energy usage. There are three general types of lighting that are widely used in the household nowadays: the (light-emitting diode) LED bulb, (compact fluorescent light) CFL bulb and the incandescent bulb. But each of them has their perks and perils [3]. Unfortunately, most of the time, people in the residential end up choosing a lamp based on it is price rather than its efficiency and this cause us to pay for our lighting more than necessary.

Today, CFLs are the most used lighting sources in the household. Using CFLs will allow people to decrease their energy consumption; also it is a good start to decrease greenhouse emissions. However, the main problem of CFLs is the mercury they contain.

When products and wastes containing mercury are improperly disposed of, mercury is released into the air, ground or water. It is persistent in the environment; it never breaks down nor goes away.

In addition to the harmful effects of mercury is that it emits Ultraviolet (UV) Radiation. This UV radiation interacts with the chemicals on the inside of the bulb to generate light. The acute and chronic effects are the normal responses of the skin to UVR; acute reactions considered will be erythema (sunburn) and vitamin D production. Skin aging and skin cancer will be discussed as those reactions produced by prolonged or repeated UVR exposure [4].

In response to the aforementioned situation where the three types of lights are compared by their efficiency and effectiveness as light sources, the proponents conducted this study. Since the CFLs are the most commonly used light bulbs in the residential and has a great compatibility to LED in terms of design and materials, the proponents conducted a study about innovating and recycling CFL into a more efficient and environment friendly LED light bulb.

2. Materials and methodology

The following materials and methods were used in this study.

2.1. Investigation phase

Data about present average usage and disposal of CFL shall be gathered. The researchers chose CFL to use and recycle because of its still huge volume of sales [5] and great compatibility to LED case. Data and specifications of LEDs and CFLs should also be collected and recorded for future reference of test and comparisons.

2.2. Design phase

The proponents have developed a circuit and designed the layout for the power supply of the SCLED light bulb adapted from the different components recycled or salvage from the busted CFL. Figure 3 shows the developed PCB layout for the power supply. The proponents also designed the layout for the LED strips as shown in figure 4.

IOP Conf. Series: Earth and Environmental Science **114** (2018) 012003 doi:10.1088/1755-1315/114/1/012003

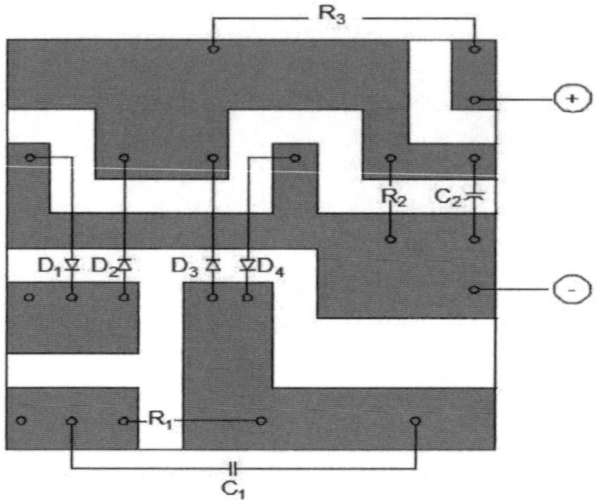

Figure 3. Printed circuit board (PCB) layout for power supply

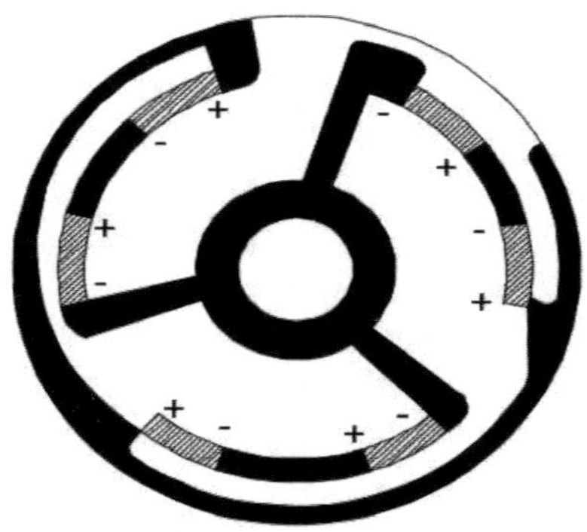

Figure 4. PCB layout for LED strips

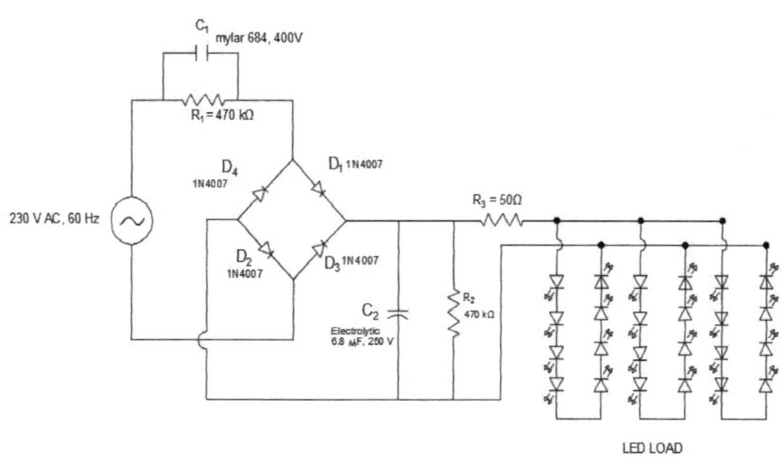

Figure 5. Schematic diagram

Figure 5 shows the schematic diagram of the circuit used and applied by the proponents in developing the SCLED bulb. The values and components are based on the availability in the busted CFL and the circuitry composition of LED lamps. The circuit values and components may vary because of the different circuitry and components used by different bulb manufacturers. The circuit flow and design for SCLED remains the same.

2.3. Construction phase
This part includes the actual design and position of power supply, the design and construction of the case of prototype, and the actual fabrication of the prototype. Figure 6 demonstrates the step by step procedure in constructing the prototype; from gathering the materials and recycling to fixing the parts and finalizing the prototype.

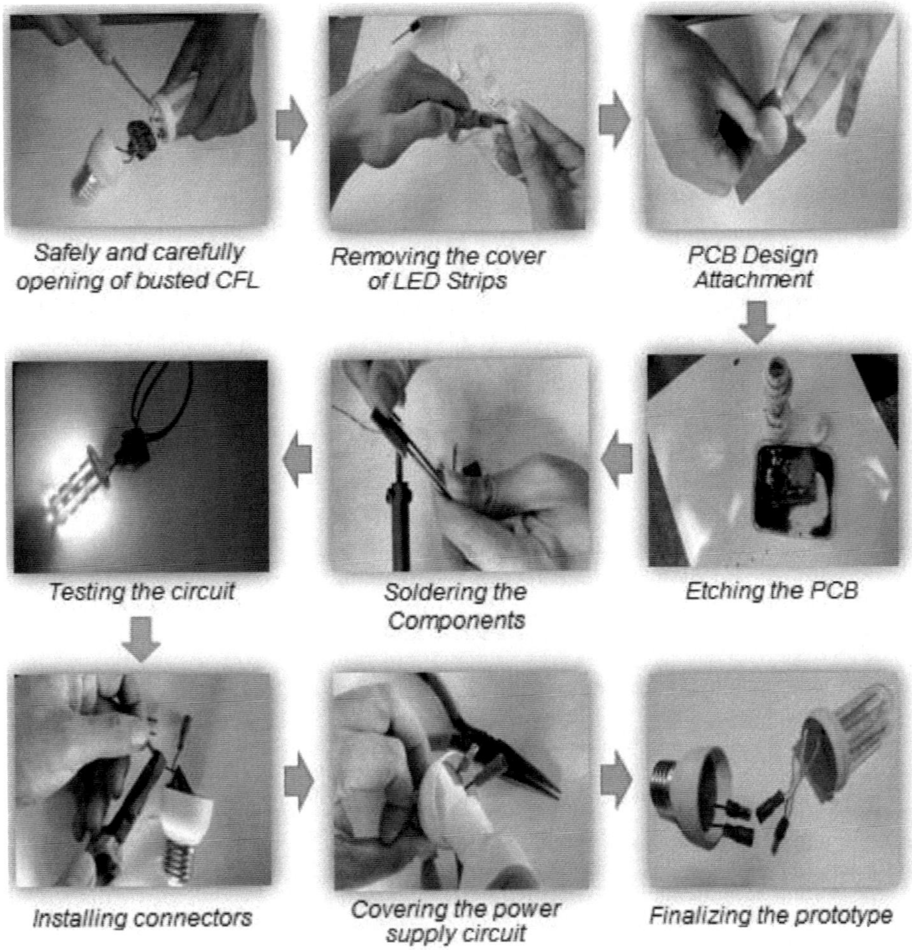

Figure 6. Construction procedure

2.4. Testing using light meter
Light meter is an instrument used to measure illuminance. The measurements were recorded from one-meter vertical distance and one-meter horizontal distance. The researchers recorded and tabulated the data and measurements every hour for a whole day. The researchers also measured the illuminance of the bulbs from one-meter distance with the light bulbs being closed and open again every 5 minutes in a one-hour continuous operation.

2.5. Testing using infrared thermometer

Infrared thermometer is a device used to measure the temperature of a certain area using infrared. The researchers recorded and tabulated the data and measurements every one hour for a whole day. The data and measurements were compared in natural room temperature and the actual room temperature.

3. Results and discussion

3.1. Illumination tests

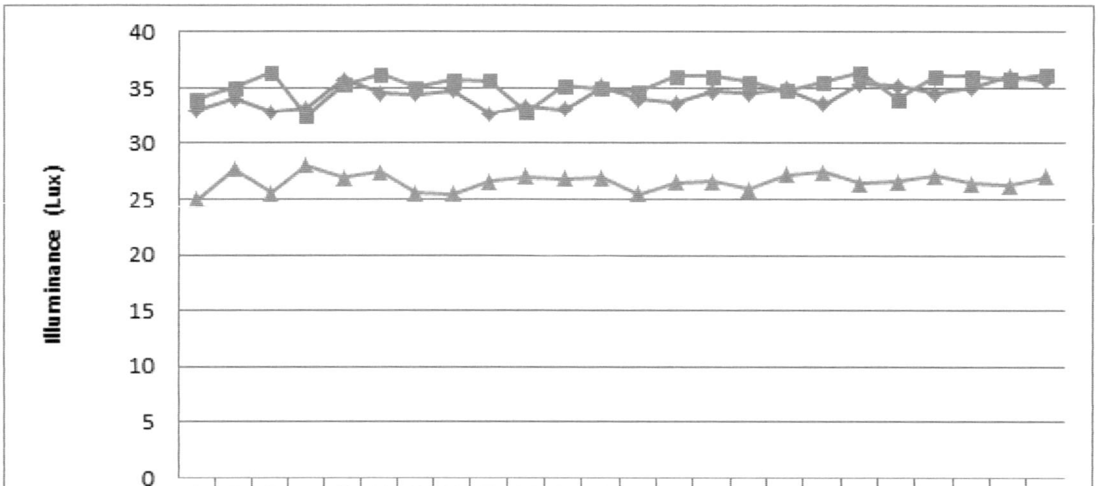

	1	2	3	4	5	6	7	8	9	10	11	12	13	14	15	16	17	18	19	20	21	22	23	24
SCLED	33	34	33	33	36	35	34	35	33	33	33	35	34	34	35	35	35	34	35	35	35	35	36	36
LED	34	35	36	33	35	36	35	36	36	33	35	35	35	36	36	36	35	36	36	34	36	36	36	36
CFL	25	28	26	28	27	27	26	26	27	27	27	27	26	27	27	26	27	27	26	27	27	26	26	27

Figure 7. Lux measurement from one meter direct vertical distance

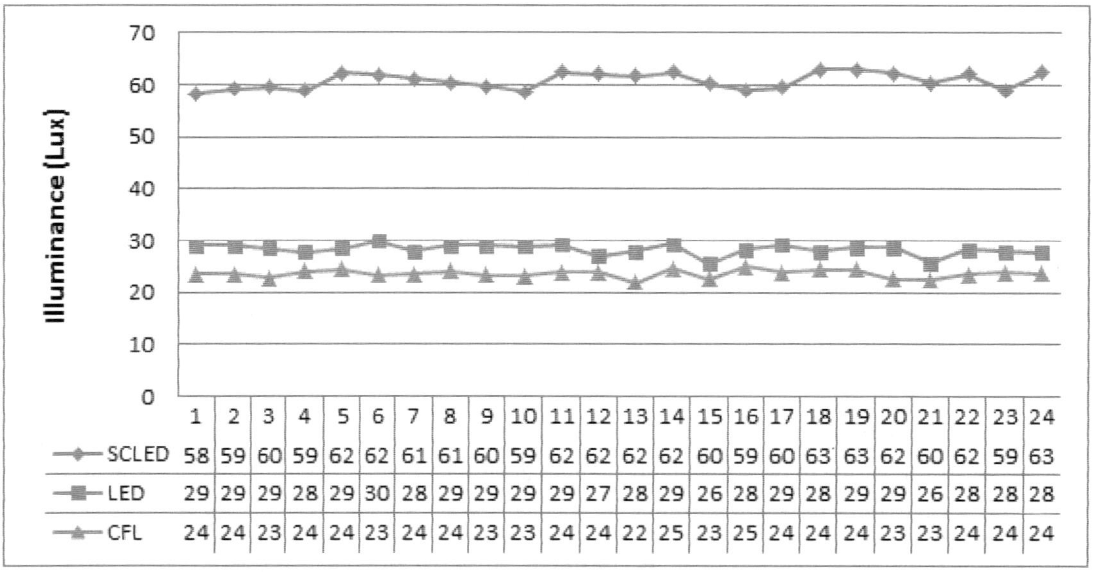

	1	2	3	4	5	6	7	8	9	10	11	12	13	14	15	16	17	18	19	20	21	22	23	24
SCLED	58	59	60	59	62	62	61	61	60	59	62	62	62	62	60	59	60	63	63	62	60	62	59	63
LED	29	29	29	28	29	30	28	29	29	29	29	27	28	29	26	28	29	28	29	29	26	28	28	28
CFL	24	24	23	24	24	23	24	24	23	23	24	24	22	25	23	25	24	24	24	23	23	24	24	24

Figure 8. Lux measurement from one meter direct horizontal distance

Figure 7 shows the data gathered using a lux meter to measure the illuminance of SCLED versus the

IOP Conf. Series: Earth and Environmental Science **114** (2018) 012003 doi:10.1088/1755-1315/114/1/012003

illuminance of a commercial LED and CFL in a 24-hour operation. It is measured every hour from directly one meter below the light source. The data gathered using lux meter from one meter horizontal distance shows a great difference between SCLED and CFL as depicted in figure 8.

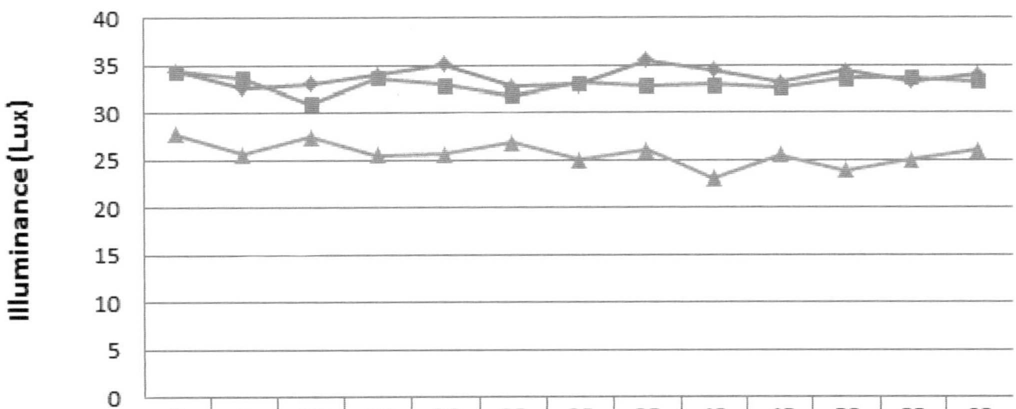

	0	5	10	15	20	25	30	35	40	45	50	55	60
SCLED	34.4	32.7	33.1	34	35.1	32.8	33	35.5	34.5	33.2	34.4	33.3	34
LED	34.4	33.7	31	33.7	33	31.8	33.2	32.9	33	32.7	33.6	33.7	33.3
CFL	27.7	25.6	27.4	25.5	25.6	26.8	25	26	23	25.5	23.9	25	26

Figure 9. Lux measurement from one meter direct horizontal distance

Figure 9 illustrates the data gathered using a lux meter to measure the consistency of the illuminance of SCLED versus the illuminance of a commercial LED and CFL in a one-hour operation. The light is switched on and off every five minutes then measured the illuminance using the lux meter.

3.2. Temperature test

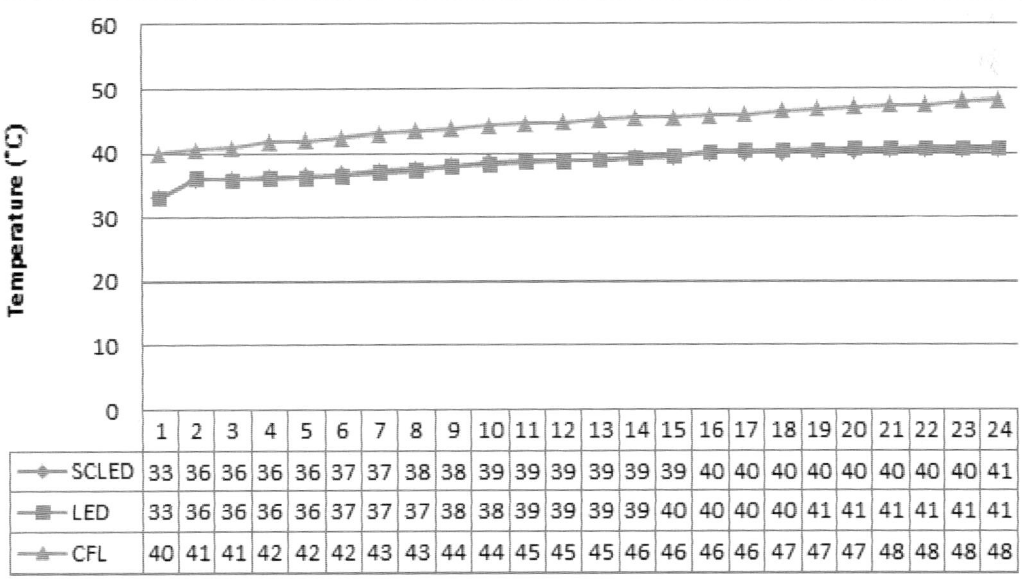

	1	2	3	4	5	6	7	8	9	10	11	12	13	14	15	16	17	18	19	20	21	22	23	24
SCLED	33	36	36	36	36	37	37	38	38	39	39	39	39	39	39	40	40	40	40	40	40	40	40	41
LED	33	36	36	36	36	37	37	37	38	38	39	39	39	39	40	40	40	40	41	41	41	41	41	41
CFL	40	41	41	42	42	42	43	43	44	44	45	45	45	46	46	46	46	47	47	47	48	48	48	48

Figure 10. Lux measurement from one meter direct horizontal distance

Operating temperature as one of the factors affecting the quality of light bulbs was also observed in this study. Figure 10 shows the temperature curve of SCLED versus the commercial LED and CFL. The data were recorded every hour for a whole day continuous operation.

3.3. Energy consumption test

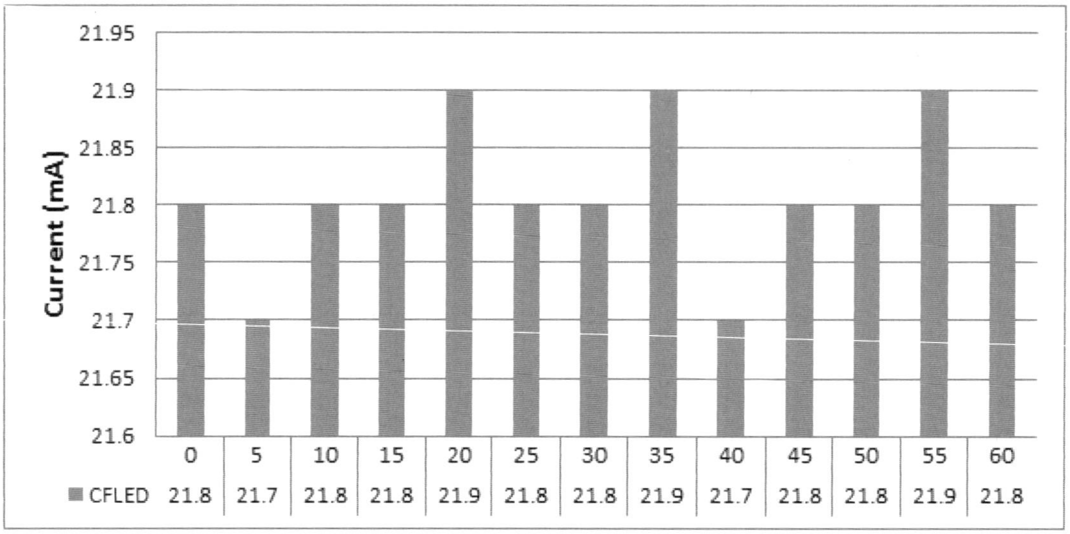

Figure 11. Current measurement

Figure 11 depicts the current drawn by the prototype in a 230V, 60Hz line. It is recorded every five minutes in a one-hour continuous operation.

3.4. Data Analysis and Computation

$$\text{Iave} = \frac{\sum current\ measurements}{total\ number\ of\ test\ points} \tag{1}$$

$$\text{Iave} = \frac{2(21.7mA)+8(21.8mA)+3(21.9mA)}{13}$$

$$\text{Iave} = 21.8077\ \text{mA}$$

$$P = VI \tag{2}$$

$$P = (230\text{V})\ (21.8077\ \text{mA})$$

$$P = 5.01577\ \text{watts} \sim 5\ \text{watts}$$

Table 1. Prototype savings vs. CFL

	SCLED	CFL
Life Span	Approx. 25,000 -50,000 hours	2,000-8,000 hours
Wattage	5 watts	15 watts
Electric Rate	8php/kwh	8php/kwh
Operating hours	10 hours/day	10 hours/day
Charge per day	0.4php	1.2php
Charge per month	12php	36php
Charge per year	146php	438php
Yearly Savings	292php	

The prototype is observed to be much more economical than the commercial CFL. Table 1 shows the advantages as well as the computed yearly savings in using the prototype compared to the commercial CFL. Considering the life span of the light bulbs, the prototype is at least three times as much as durable as the commercial CFL. Using same electric rate and number of hours operated in a day, the data were computed and noticed to have a great difference in terms of operating cost per year.

4. Conclusion

SCLEDs have a lot of advantages compared to CFL, incandescent and other lighting lamps in residential and other applications. The power consumption is lower than other bulbs, therefore allowing customers to pay little amount of money in lighting energy consumption. It has longer life span compared to others, thus, making it more economical in the long run. It has a segmented feature that makes it more economical. If the LED strips are damaged, you can easily remove it from the SCLED case and replace it with new one without changing the other parts. Same thing will be done if other parts of the SCLED are damaged; user does not need to replace the whole bulb, instead, just replace the damaged part. SCLED is composed of parts and components from busted CFL, making it cheaper compared to the commercial LED. On the other hand, commercial CFLs may be cheaper or almost at the same price. However, the power consumption throughout is lower compared to those CFLs. To sum it up, in the long run, SCLED is cheaper and has a higher efficiency. It can contribute to the aspect of power consumption in the whole world scenario. This project has proven that LED bulb can be made out of recycled CFL. It only uses recycled components and plastic casing from the busted CFLs that can contribute in the reduction of waste materials. In addition, SCLEDs do not contain harmful chemicals such as mercury that can pollute the environment.

5. Acknowledgment

Our grateful appreciation to Dr. Gregorio J. Rodis, the University President, for granting us the opportunity to work on this study and for his endless quest to enrich research culture in the university especially among the faculty members of BPSU. The study would not be completed without the help of BSEE-5 students especially those who helped in the construction of the prototype. Thank you to, Renee Jane San Pedro-Ponce, the proofreader, and Altroye Manlangit, Jon Francis Pagador, Mark Anthony Mayuyo, Mia Sharifa Mesina, and Monalie Pujante for aiding the proponents in carrying out this study. Most of all, we express our heartfelt gratitude to Almighty God for His unwavering goodness, guidance and wisdom that He had provided throughout this research.

6. References

[1] Philippine Statistic Authority (2013) Energy is the most common source of energy used by households.
[2] Gallachoir B, O'Leary F and Howley M (2008) Energy in the Residential Sector.
[3] Francisco K B (2015) Difference between Incandescent, Compact Fluorescent Light and LED.
[4] Havas M (2008) Health Concerns associated with Energy Efficient Lighting and their Electromagnetic Emissions
[5] Berkeley L (2011) Energy-Efficient Compact Fluorescent Table Lamp

IOP Conf. Series: Earth and Environmental Science **114** (2018) 012004 doi:10.1088/1755-1315/114/1/012004

Heat radiation approach for harnessing heat of the cook stove to generate electricity for lighting system and charging of mobile phone

Rodrigo C. Muñoz, Jr, [1] Chad Deo G. Manansala [2]

[1, 2]Electronics Engineering Department, Bataan Peninsula State University, Balanga City, Bataan, Philippines

Abstract. This study is based on the potential of thermoelectric coupling such as the thermoelectric cooler module. A thermoelectric cooler converts the heat coming from the cook stove into electricity and store in a battery. A dc-dc boost converter will be used to produce enough voltage to light a minimum house dwelling or charge phone battery. This device will be helpful to those that faces a problem on electricity especially in the isolated areas. The study aims (1) to harness heat from the cook stove up to 110 0C; (2) To automatically cool-off the system to protect the thermoelectric cooler from damage due to excessive heat using an electronic solenoid; (3) To store energy harnessed in the battery; (4) To amplify the output voltages of the battery using DC to DC boost converter for lighting system and charging of mobile phone battery. From various tests conducted, it can fully charge a mobile phone in 3 hours observing the unit's battery voltage drop from 4.06V to 3.98V. In the testing it used different orientation of steel rod by conduction to transfer heat and by radiation through tubular steel with its different dimensions. Most recent testing proved that the 2x2x9 tubular steel by radiation had the best result. The temperature reached more than a hundred degree Celsius that met the objective. The test resulted of boosting the voltage of the battery output from 3.7V to 4.96V on the average. The boosted voltage decrease as the system's cool-off mechanism operated when the temperature reached above 110 degree Celsius decreasing output voltage to 0.8V resulting the boosted voltage to drop to zero. Therefore, the proponents concluded that heat waste can be converted to electrical energy by harnessing heat through radiation, with the help of TEC that generates voltage for lighting and can be boosted to be used for mobile charging. Furthermore, the study proved that the excess heat can damaged the TEC which was prevented by using of cooling-off mechanism, making it more useful for longer time.

1. Introduction

The foundation of this research is the study of Alan Pnakovich in the year 2009 of American Heritage School, Plantation, Florida about the Seebeck and Peltier Thermoelectric effects, the two junctions that are made between two different conductive metals with different temperatures will create a voltage from the temperature difference of the two junctions [1]. Thermoelectric coupling process is literally associated with thermal and electrical phenomena. It can directly convert thermal energy into electrical energy or vice versa. [2]. The TEC has a parallel thermal connection, hot side and the cold side. It will serve as the input in the thermoelectric cooler. A low voltage of the order of tens of micro voltage per kelvin is generated. This phenomenon is called Seebeck effect [3]. The power generation of thermoelectric devices are low efficiency at heat-to-electricity, which typically near to 5% thus the thermoelectric device is limited to output only a low voltage.

This research is motivated by the study of O'Shaughnessy in his study about the "Small scale electricity generation from a portable biomass cook stove that thermoelectric module is capable of delivering small amounts of off-grid electricity". A single thermoelectric module is utilized to convert a small portion of heat from the stove to electricity [4]. The research started the initial study in 2011 with a project feasibility study in "Direct Conversion of Heat to Electrical Energy" [5]. The researcher's objective was to supply electricity to the isolated areas using the heat from the cook stove that is harnessed using the metal rod using the thermoelectric generator (TEG). In the study of Daniel Champier and his team of Thermoelectric Generator Incorporated in a Multifunction Wood Stove, a fixed water tank in the stove as the exchanger is being attached in the cold side of the TEC for the coolant [6]. This type of coolant helps the TEC to produce greater power compared to other type of coolants. Liquid in cold side can sustain a very large differential temperature even with an elevated ambient temperature. This was adapted by this study. The research also implemented boost converters used to convert the lower input voltage into higher output voltage [7]. The voltage regulator process the regulating of the voltage of the TEC before storing it in a battery. Voltage amplification step-up the battery voltage to supply the electronic module, temperature sensor and to charge a mobile phone battery. The electronic module of the system, consist of a battery storage where the output voltage from the coupling unit is simply stored. In this study, the proponent used secondary batteries (rechargeable) with a specification of 5000mAh, 3.7 V single battery.

2. Significance of the study

There are many kinds of waste heat, and most of it just disappears in the air and is not useful in any kind of way. Waste heat from a cook stove is abundant, and every household uses it. Turning waste heat into electricity can be a great help to each community in areas experiencing power shortage with special notation of climate change.

According to the law of conservation of energy, energy can neither created nor destroy. It can only convert from one form to another [8]. Thermoelectric module converts heat waste to electrical energy and store in an appropriate battery bank by implementing a DC charger circuit in order to be a useable source of electricity. A single thermoelectric module is utilized to convert a small portion of heat from the stove to electricity. A thermoelectric module has an ability to generate power and provide a source of heating and cooling. Each of which is consist of p-type and n-type semiconductor that is connected thermally in parallel and electrically in series. The development of a simple and efficient means of converting waste heat to electricity would be a good alternative source of energy that can help the community's problem of power shortage.

3. Methodology

The researchers studied the concepts and design of the heat to electricity electronic module needed to convert the bursting heat energy to electrical energy. The cooling-off mechanism was designed and installed to a heat source to protect TEC from excessive temperature required. Metallic rod is attached in a way it is configured to conduct heat. The temperature sensor on the metal plate (hot side of the thermoelectric cooler) will sense if the heat flowing is exceeding in the set temperature. It will trigger the solenoid lock to actuate. The thermoelectric cooler (TEC) is stand-off between the metal plate and galvanized iron container to have temperature difference across the hot side and the cold side to convert the heat to electricity. The output voltage from the thermoelectric coupling unit will be the inputted to the circuit of a DC to DC boost converter to amplify the generated power of the thermoelectric coupling unit to attain the output needed to charge the battery. The generated electricity from the thermoelectric cooler unit will be stored in the battery storage with a specification of 2300 mAh, 3.7 volts rechargeable battery. It will undergo voltage amplification process to charge a mobile phone battery of 5 Volts.

4. Test results

4.1. Harnessing heat from cook stove and storing energy harnessed in the battery

Harnessing heat from the cook stove in terms of heat temperature were monitored every 1 hour of cooking time. The following figures show the testing of the proponents between conduction and radiation.

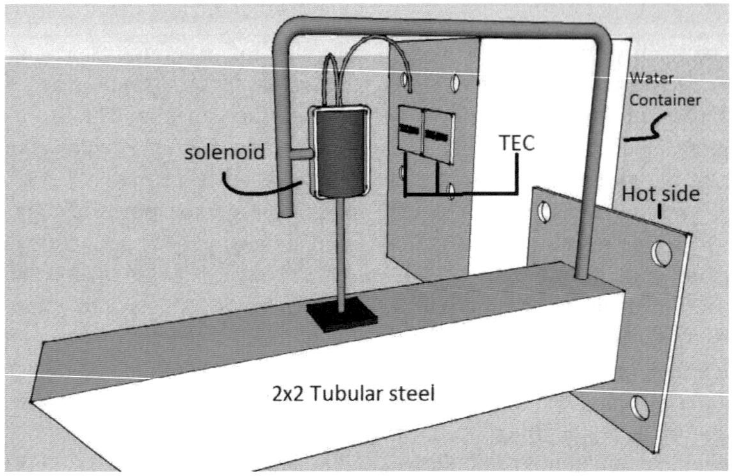

Figure 1. Developed prototype

In heat harnessing, evaluation was conducted in stove connected with metals that transfers heat through conduction. Figure 2 show the thermal conduction of stove within 60 minutes of testing. As indicated, the first set up used solid metal that does not reach 100° which does not meet the objective. There is an instance of a sudden dropped which was caused by the lack of charcoal to produce heat.

For the figure 3 used radiation method, it indicated the five (5) sizes of tubular used for radiation. For the 2x2x11 tubular, the heat harnessing was good, for it continuously increases until it reach 100°C then cool-off which caused the temperature to decreased down to below 90°, then, when the cooling-off mechanism changed actuation, the temperature increase again. Compared to other sizes of the tubular used; only the 2x2x11 has reached 100°. The two figures indicated that using radiation to harness heat is better than harnessing heat via conduction. Figure 4 compared the heat transfer via radiation and conduction. Copper was used as a thermal bridge in conduction graph. Knowing copper that has high thermal conductivity and TEG is used capable of handling 800°C. On the other hand, by radiation, tubular was used as thermal bridge and TEC as its module. Tubular is much cheaper than copper and TEG is not available locally in the Philippines. TEC can only handle 110°C and may be damaged if the temperature exceeds that limit. Tubular is enough for the TEC to function. In figure 5, it was shown that the charging of battery unit continuously increases until it fully charged to 4.2V. It took almost 3 hours to fully charge the battery of the unit.

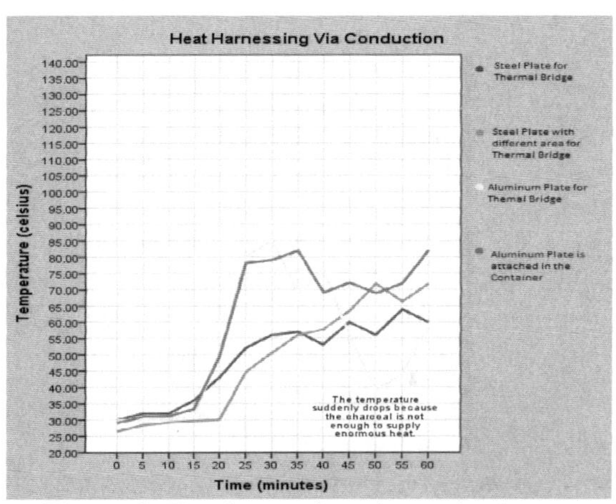

Figure 2. Heat harnessing via conduction

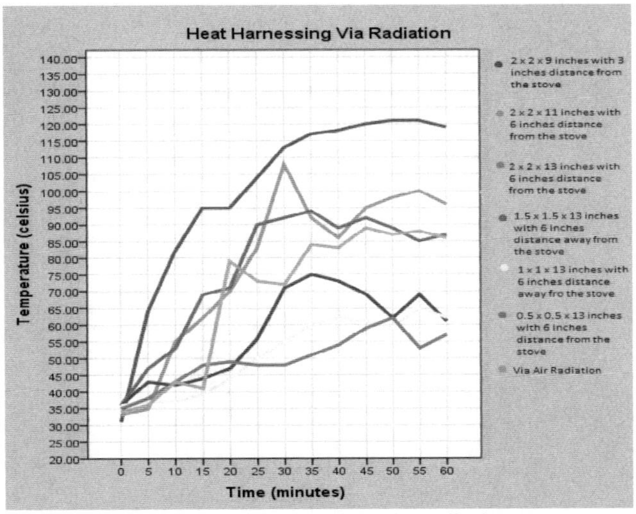

Figure 3. Heat harnessing via radiation

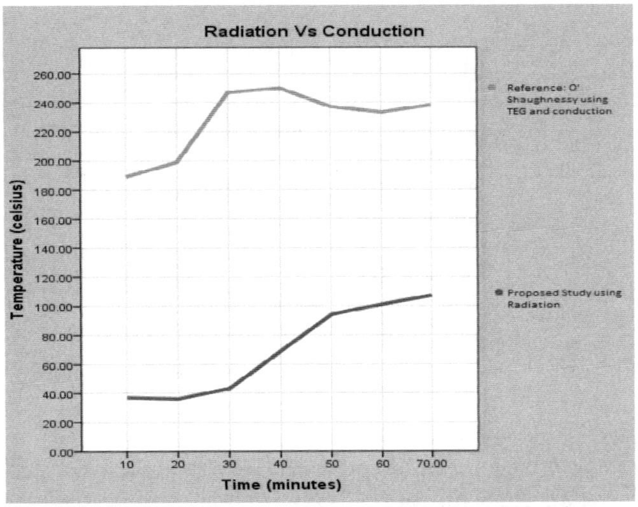

Figure 4. Comparing of radiation and conduction

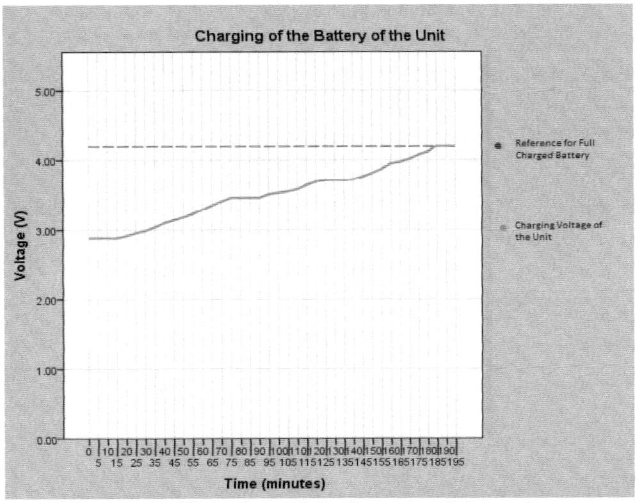

Figure 5. Charging of the battery of the unit

4.2. Monitoring the "cooling-off" mechanism

To monitor the "cooling-off" mechanism, the objective is to cool-off the system not exceeding 110°C. From the several testing conducted, the histogram in Figure 6a and Figure 6b shows that the sensor can sense the maximum temp of 110 °C and actuated the solenoid to cool-off the system.

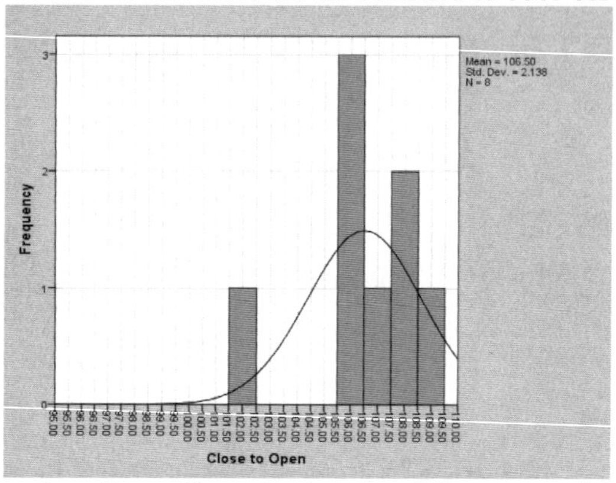

Figure 6a. Monitoring of cooling-off mechanism for different condition (close to open)

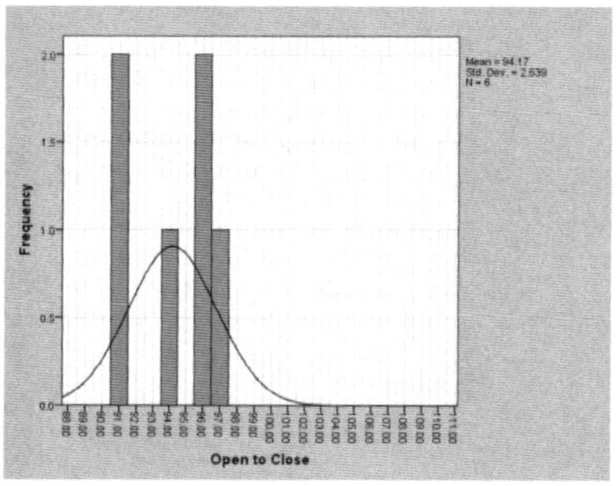

Figure 6b. Monitoring of cooling-off mechanism for different condition (close to open)

4.3. Amplifying the voltage of the battery and effectiveness fo the device

In figure 7, with the input voltage of 3.7V to the battery the highest boosted voltage is 4.96V. The required charging voltage for the phone of 5V is also indicated in the figure. Shown also in Figure 8 are the boosted voltage required to charge the unit. It decreased because it cools-off when the temperature decreases resulted the output voltage of the TEC to drop below 0.8V, the boosted voltage will then result to 0V. The required voltage to charge the mobile unit is 3.7V.

Figure 9 shows the linear relationship of time and battery charge percentage of the mobile phone. This figure indicates that the system unit can charge a battery of mobile phone with a specification of 5V and 2300 mAH. In several testing, charging time of the mobile phone battery required 3 hours to fully charge.

In figure 10, comparing the light intensity with the other study, the light intensity using a sun king lantern lasted for five hours with a 2 watts light (25lux). Comparing the previous study with the proposed method, the 1 watt LED lasted for three hours of fixed intensity of 25 lux and continuously decrease as it dropped to 15 lux at the end of 5 hours.

IOP Conf. Series: Earth and Environmental Science **114** (2018) 012004 doi:10.1088/1755-1315/114/1/012004

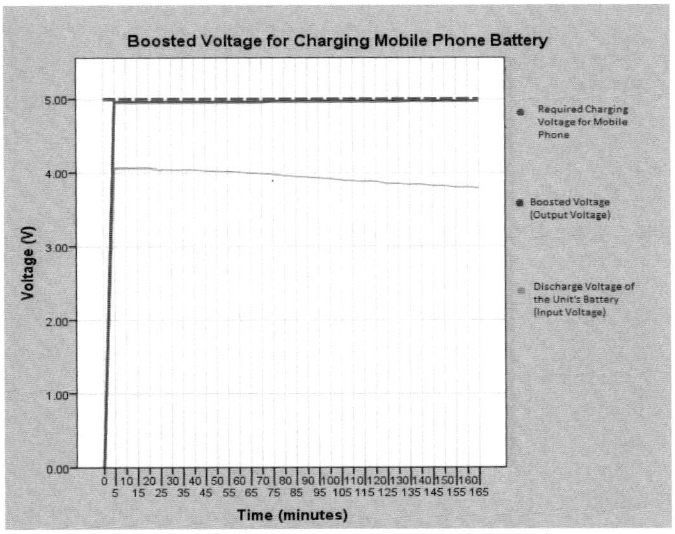

Figure 7. Boosted battery voltage of the unit

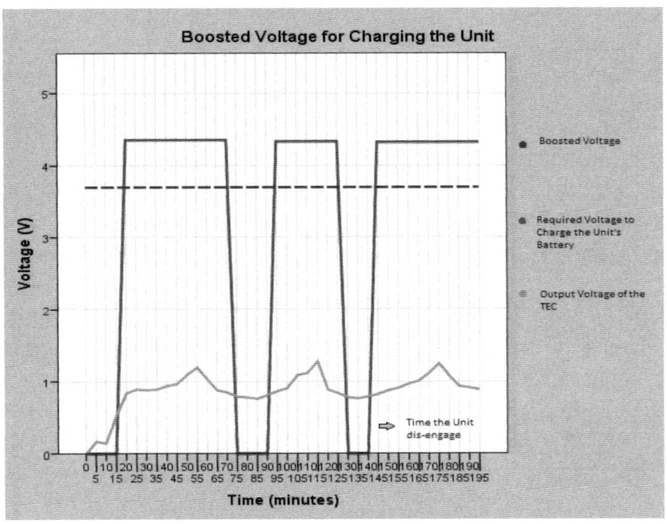

Figure 8. Boosted voltage for charging the unit

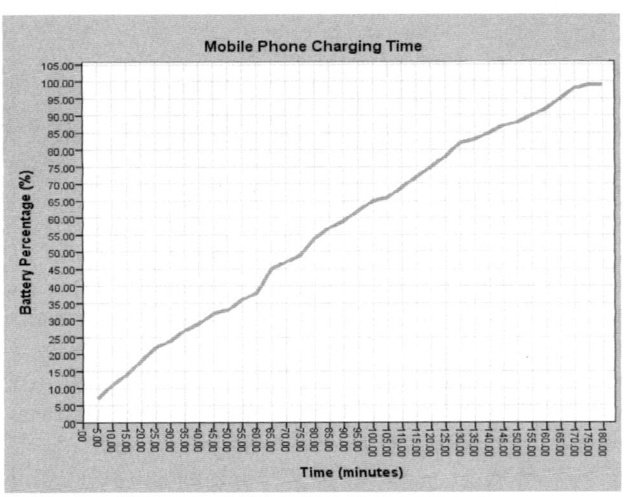

Figure 9. Percent charging mobile phone battery

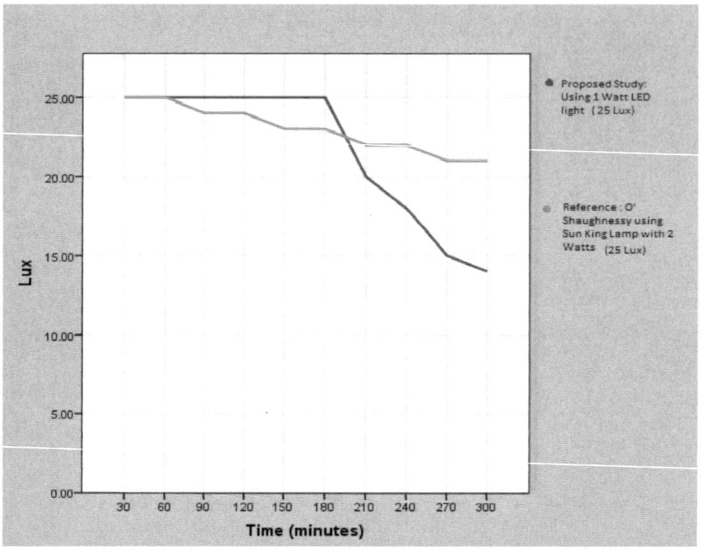

Figure 10. Light intensity produced of fully charged battery

5. Conclusion

Based on the various dataa and results of the study, the proponents concluded that the use of radiation as a way for heat harnessing performed better than that of conduction. By using TEC instead of TEG can produced equal performance with comparable application taking into account the use of "cooling-off" mechanism to prevent the TEC from excessive heat which may cause damage or affects the performance. Safekeeping the TEC from damage made it more useful for longer time.

6. Acknowledgment

Based on the various dataa and results of the study, the proponents concluded that the use of radiation as a way for heat harnessing performed better than that of conduction. By using TEC instead of TEG can produced equal performance with comparable application taking into account the use of "cooling-off" mechanism to prevent the TEC from excessive heat which may cause damage or affects the performance. Safekeeping the TEC from damage made it more useful for longer time.

7. References

[1] Pnakovich, A. A study of the Seebeck and Peltier Thermoelectric Effects. American Heritage School, Plantation, Florida. Retrieved from https://www.scribd.com/doc/36322318/A-Study-of-the-Seebeck-and-Peltier-Thermoelectric-Effects. 2009.

[2] Yan David. (2011). Modeling and Application of a Thermoelectric Generator. Undergraduate Thesis Department of Electrical and Computer Engineering. University of Toronto. Retrieved from https://tspace.library.utoronto.ca/bitstream/1807/31648/3/Yan_David_201111_MASc_thesis.pdfR. Dewri, and N. Chakraborti. Simulating recrystallization through cellular automata and genetic algorithms. Modelling Simul. Mater. Sci. Eng. 2005, 13 (3): 173-183.

[3] Bitschi. A. Modeling of Thermoelectic Devices for Electric Power Generation. Doctoral Thesis. Swiss Federal Institute of Technology Zurich, Zurich, Switzerland. Retrieved from http://e-collection.library.ethz.ch/eserv/eth:470/eth-470-02.pdf. 2009

[4] O'Shaughnessy. S.M.M., Deasy M.J., Kinsella C.E. Doyle, J.V., RobinsonA.J. Small Scale electricity generation from a portable biomass cook stove: Prototype design and preliminary results. Department of Mechanical & Manufacturing Engineering, Parsons Building, Trinity College, Dublin 2, Ireland. Retrieved from http://www.tara.tcd.ie/xmlui/bitstream/handle/2262/67185/small%20scale.pdf?sequence=1&is

Allowed=y. 2012.

[5] Muñoz, R. Jr., Apostol. J.P., Banzon M.J.V., Cahapay N. S., Lopez M.M., Navarro C.A. (2011). A Project Feasibility on Direct Conversion of Heat to Electrical Energy. Undergraduate Thesis. Bataan Peninsula State University, Balanga City of Bataan, Philippines, 2011.

[6] Champier D., Bedecarrats J.P., Kouskou T.,Rivaletto M., Strub F., Pifnolet P. Study of a Thermoelectric Generator Incorporated in a Multifunction Wood Stove. Retrieved from https://hal.archives-ouvertes.fr/hal-00727162/document. , 2011.

[7] Agrawal, J. P. Power Electronics Systems: Theory and Design. Prentice-Hall, Upper Saddle River, NJ, chap. 6, 2001.

[8] Joshi, D.M. A book of Science and Technology: Living Science Physics 9. Ratna Sagar P. Ltd. 2007, p 130.

Experimental Research on Optimizing Inlet Airflow of Wet Cooling Towers under Crosswind Conditions

You Liang Chen[1], Yong Feng Shi[1], Jian Gang Hao[1], Hao Chang[1], Feng Zhong Sun[2]

[1]Huadian Electric Power Research Institute, No. 2 Xiyuan Ninth Road, Hangzhou 310030, China
[2]Shandong University, No. 17923 Jingshi Road, Jinan 250061, China
E-mail: cyl_sdu@163.com

Abstract. A new approach of installing air deflectors around tower inlet circumferentially was proposed to optimize the inlet airflow and reduce the adverse effect of crosswinds on the thermal performance of natural draft wet cooling towers (NDWCT). And inlet airflow uniformity coefficient was defined to analyze the uniformity of circumferential inlet airflow quantitatively. Then the effect of air deflectors on the NDWCT performance was investigated experimentally. By contrast between inlet air flow rate and cooling efficiency, it has been found that crosswinds not only decrease the inlet air flow rate, but also reduce the uniformity of inlet airflow, which reduce NDWCT performance jointly. After installing air deflectors, the inlet air flow rate and uniformity coefficient increase, the uniformity of heat and mass transfer increases correspondingly, which improve the cooling performance. In addition, analysis on Lewis factor demonstrates that the inlet airflow optimization has more enhancement of heat transfer than mass transfer, but leads to more water evaporation loss.

1. Introduction

Natural draft counter-flow wet cooling tower is widely used to cool the circulating water in thermal power plants and nuclear power plants. Circulating water from the condenser is injected from the nozzles and goes successively through spray zone, filler zone and rain zone to transfer heat and mass to the air, and then returns to the condenser. The cooling tower performance decides directly the condenser vacuum and influents the generating efficiency [1], so it is of great importance to research and improve the cooling performance. There are mainly three methods of cooling tower thermal calculation, namely Poppe, Merkel and e-NTU. Kloppers et al. [2] found that the results of Merkel and e-NTU are nearly the same while the accuracy of Poppe is the highest. Fisenko et al. [3] proposed a cooling tower performance mathematical model including both the droplet cooling in spray zone and the film cooling in fill zone. The deviation between the calculated results and experimental results is smaller than 3%, but the model is not applicable to frozen conditions and windy conditions. Hawlader [4] and Williamson [5] respectively used algebraic method and k-ε turbulence model to build two-dimensional axisymmetric cooling tower numerical model and study the influence of the nonuniformity of air water flow and heat and mass transfer on the cooling tower performance. Smrekar [6] measured the air flow field above the spray zone and proposed the optimal spray mode which can ensure the uniformity of the heat transfer inside tower and minimize the available energy loss. Muangnoi [7] found through the available energy analysis that the available energy provided by circulating water is more than that obtained by air due to entropy generation. The available energy loss in top of the heat and mass transfer zones is larger while bottom smaller. Other than the cooling tower structure factor, environmental factors especially the cross wind also have strong impact on cooling tower performance [8-10]. Derksen [11-13] found through wind tunnel test and numerical calculation that crosswind can strongly affected the air flow field and proposed that windbreaks should be installed on the upper air inlet to reduce the air from upwind side, thus ensure the uniformity of inlet air. According to Zhai et al. [14], the main reason of the crosswind negative effect is that crosswind changes the pressure distribution of the inlet and outlet and breaks the uniformity of inlet air, and windbreaks can weaken the negative effect of crosswind. Al-Waked et al. [15-16] used fixed velocity droplet flow to simulate the film flow in fill zone and found trough numerical calculation that the outlet water temperature is increased by 1.7 ℃ when the crosswind velocity is 7.5 m/s. The primary cause is proposed to be the nonuniformity of the air flow field caused by crosswind. Research also shows that installing porous windbreaks both inside and outside rain zone can lower the outlet water temperature by 0.5~1 ℃. Gao et al. [17] found through model experiment that when Froude number is 0.174 the cooling temperature difference and efficiency coefficient have the minimum value. Zhao et al. [18-19] built a 3D numerical model applicable to cooling towers under crosswind effects and found that crosswind increases the inlet air relative deviation, generates cross air flow and strengthens rain zone performance, meanwhile weakening heat and mass transfer in the fill zone, lowering the water temperature difference of fill zone and increasing the outlet water temperature. It's proposed that installing cross wall in rain zone will improve the tower performance. Zhou et al. [20-21] use high-accuracy Poppe model in fill zone and Discrete Phase Model in spray zone and rain zone, and obtain the 3D numerical solution of wet cooling tower. Results show that crosswind has negative effects on outlet water temperature which gets the highest value when the crosswind velocity is 6 m/s, 1.34 ℃ higher than zero crosswind conditions. Installing cross wall can lower the outlet water temperature by 0.32 ℃.

In conclusion, previous researches consider through qualitative analysis that crosswind breaks the uniformity of cooling tower inlet air flow and reduces the inlet air quantity, thus weakening the cooling performance. Besides, no better solution than windbreaks is proposed. In this paper, a kind of new method is introduced, namely air deflector. Air uniformity coefficient is defined to be the performance index evaluating the inlet air flow uniformity. Through hot model test, the impact of air flow optimization on cooling tower performance is studied, the mechanism of which is analyzed.

2. Model experiment

2.1. Test criterion

To be more valid and accurate, and guide design, operation and optimization of the real tower, the hot model test must satisfy the following similarity criteria, including the geometric, kinematic, dynamic and thermal similarity.

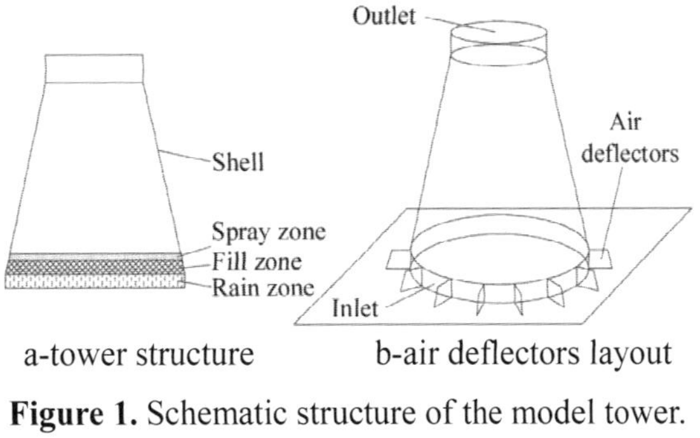

a-tower structure b-air deflectors layout

Figure 1. Schematic structure of the model tower.

2.1.1. Geometric similarity. The model NDWCT used in this paper is made with a scale of 1:100 to the prototype tower, the dimension of the prototype tower is 37 m × 68 m × 85 m (top outlet diameter × bottom diameter × height), the height of the tower inlet is 5 m, and the fill area is 3200 m^2. The dimension of the model tower is about 370 mm × 680 mm × 850 mm, and the height of the tower inlet is 50 mm. Fig. 1 shows the schematic structure of the model tower. There is an optional variable frequency fan at the top of the tower to make up for the lack of model tower pumping force. Wind tunnel experiments of the tower and its internal components were performed to ensure resistance similar with the prototype tower.

2.1.2. Kinematic similarity. The air velocity ratio of the model should be equal to that of the prototype, that is,

$$(\frac{v_{to}}{v_{cw}})_P = (\frac{v_{to}}{v_{cw}})_M \tag{1}$$

where v_{to} is the outlet air velocity, v_{cw} is the crosswinds velocity at the top outlet, P and M represent the prototype and model tower respectively.

2.1.3. Thermo-dynamic similarity. It is impossible for the model test to conform to both Reynolds criterion and Froude criterion at the same time. This experiment is hot model test, through the heat and mass transfer between the inlet air and the circulating water, making the formation of the density difference between the air inside and outside tower, resulting buoyancy driving force, and on this basis to study the thermal performance of the cooling tower under various crosswind conditions. In this model test, the driving force of buoyancy and the inertial force of crosswinds are the main factors to be concerned, while the viscous force is less important. Therefore, it is the density Froude number similarity to be satisfied, that is,

$$Fr_\Delta = (\frac{v_{to}}{\sqrt{\frac{\Delta\rho_a g H_e}{\rho_i}}})_P = (\frac{v_{to}}{\sqrt{\frac{\Delta\rho_a g H_e}{\rho_i}}})_M \tag{2}$$

where Fr_Δ is density Froude number, $\Delta\rho_a$ is the density difference between the inlet and outlet air, ρ_i is the density of the inlet air, g is the acceleration of gravity, H_e is the effective height.

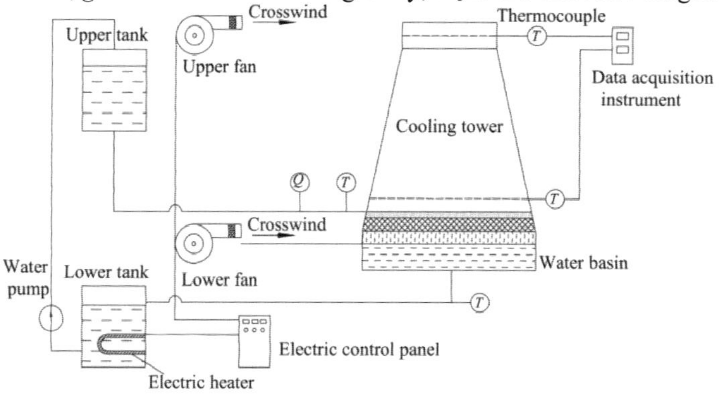

Figure 2. Schematic diagram of the hot model test system

2.2. Experimental equipments

The entire cooling tower models are made of transparent plexiglass, which is shown in Fig. 2. In the Experiment-performed, firstly, circulating water in the heating tank is heated to a setting temperature, secondly, the water is pumped into the buffer tank by using the circulating water pump. Circulating water will automatically flow into the cooling tower, then through spray zone, fill zone and rain zone in sequence and have convective heat and mass transferred with the air from the bottom to top. The cooled circulating water falls into the water basin and flows into the heating tank, a circuit is completed.

In the hot model test, air deflectors are installed at the tower inlet, which are right-angled trapezoid with thickness of 1mm, 36 pieces in total, uniformly distributed along the circumference of the tower inlet. adjacent air deflectors are spaced by 10°, and its detailed dimensions and layout are as shown in Fig. 3. In the following text, BO and AO represent before and after the inlet airflow optimization condition.

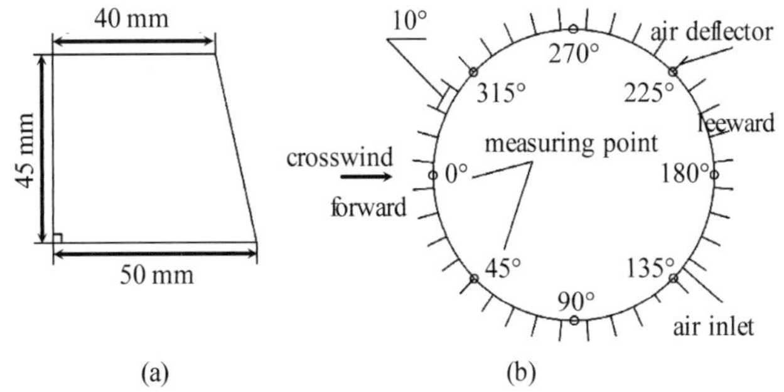

Figure 3. Arrangement mode for air deflectors. (a) air deflector size, (b) distribution of air deflectors and inlet air velocity measuring points

3. Experiment content

In the experiment, the tower inlet air velocity is measured to quantitatively analyze the inlet air uniformity along the circumference of the cooling tower. The layout of eight measuring points is as shown in Fig. 3(b) and spaced by 45°, all of which are in a half height of the tower inlet. At the same time, temperatures of the circulating water and air in and out of the tower in different side before and after optimization are measured to get various performance parameters of the cooling tower under various operating conditions, then the difference of inlet air of the cooling tower and thermal performance parameters are analyzed to clarify the effects of crosswind and air deflectors on the performance of the cooling tower.

The circulating water temperature is set as 40 ℃, and circulating water flow rate is 8 L/min. Circulating water temperature can be set by an electronic control panel, its volume can be adjusted through regulating valve on the circulating water line. An induced draft fan is installed away from the experiment table to get the hot and humid air out of the tower, in order to keep constant temperature and humidity. Due to the influence of crosswind is mainly concentrated in the air inlet and outlet of cooling tower [10, 14], frequency conversion fans are on the height of tower inlet and outlet to simulate environment influence on the cooling tower. According to the dynamic similarity criterion, crosswind velocity of the hot model test should be 1/10 of the actual velocity, six state of crosswind velocity (0.0, 0.2, 0.4, 0.0, 0.2 and 1.0 m/s) are selected in the lower fan, measured in the half height of windward side of the cooling tower (practice for z = 2.5 m height). The higher fan crosswind velocity (practice for z = 85 m) should be determined according to the formula depicting the natural crosswind velocity distribution above the ground [15], that is,

$$\frac{v_{cw}}{v_{cw,ref}} = (\frac{z}{z_{ref}})^{0.2} \tag{3}$$

where $v_{cw,ref}$ is a reference crosswind velocity at the reference height z_{ref} = 10 m. The crosswind velocity in the higher fan is twice as the lower fan, and the corresponding crosswind velocity in the higher is 0.0, 0.4, 0.8, 1.2, 1.6 and 2.0 m/s. In the following text, the crosswind velocity is indicated by the higher fan crosswind velocity. More cross-section flow field tests guarantee the reliability of the results. The test of flow field in the section shows that the model velocity have roughly the same with actual conditions, which ensured the reliability of the result.

IOP Conf. Series: Earth and Environmental Science **114** (2018) 012005 doi:10.1088/1755-1315/114/1/012005

In the experiment, atmosphere pressure, dry-and-wet-bulb temperature, wind velocity in the side and inlet, circulating water, the temperature of the circulating water and air in and out of the tower, etc. are measured. All of these monitored parameters and measuring instruments are listed in Table 1.

Table 1. Monitored parameters and measuring instruments

Item	Measuring instrument	Accuracy
Atmospheric pressure	Hot-wire manometer (KA31)	0.01 kPa
Crosswind and inlet wind velocity	Hot-wire anemoscope (KA31)	0.01 m/s
Inlet dry and wet bulb temperature	Psychrometer	0.1 ℃
Outlet air temperature	Copper-constantan thermocouple	0.1 ℃
Inlet and outlet water temperature	Mercury thermometer	0.1 ℃
Inlet and outlet air humidity	Hygrometer (HI8564)	0.1%
Circulating water flow rate	Variable area flow meter	0.01 L/min

4. Results and Discussion

4.1. Cooling efficiency

Cooling efficiency η is often used to indicate the thermal performance of NDWCT. η is determined from,

$$\eta = \frac{T_{wi} - T_{wo}}{T_{wi} - T_{awb}} \tag{4}$$

where T_{wi} and T_{wo} are the inlet and outlet water temperature respectively, T_{awb} is the wet bulb temperature of the inlet air.

As shown in Fig. 4(a), before the optimization of the inlet airflow, the cooling efficiency is increased at first, and then reduced by the increasing environment crosswind. The cooling efficiency drops from 15.93% to 14.41%, a decline of nearly 10 %, when v_{cw} increases from 0 to 0.6 m/s, η rebounds when v_{cw} continues to increase. After optimizing the air intake with wind deflectors, the overall trend of cooling efficiency is similar to that before optimization, but has been significantly weakened by the crosswind. Compared with optimized air intake, the cooling efficiency of different crosswind conditions improves significantly, its biggest decline is 4.27% by the influence of crosswind. Before and after optimization, the change of the cooling efficiency is determined by the intake performance of the cooling tower.

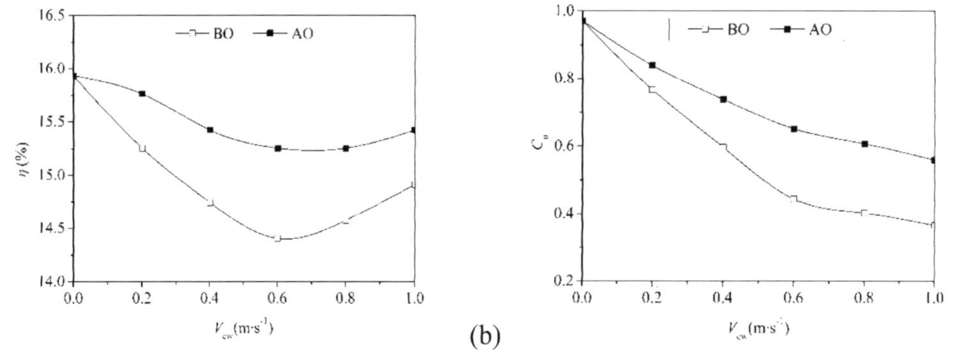

(a) (b)

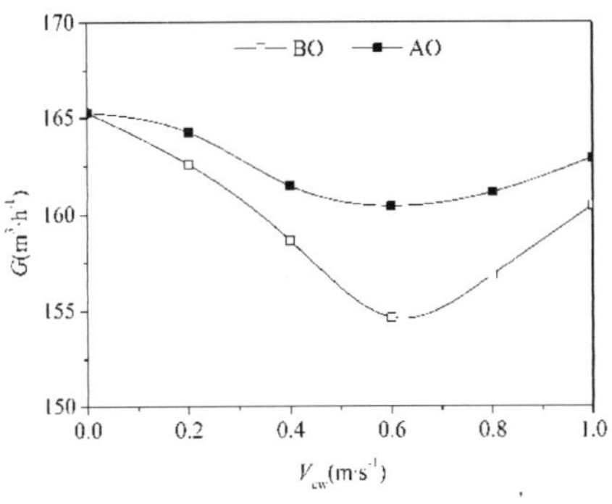

(c)

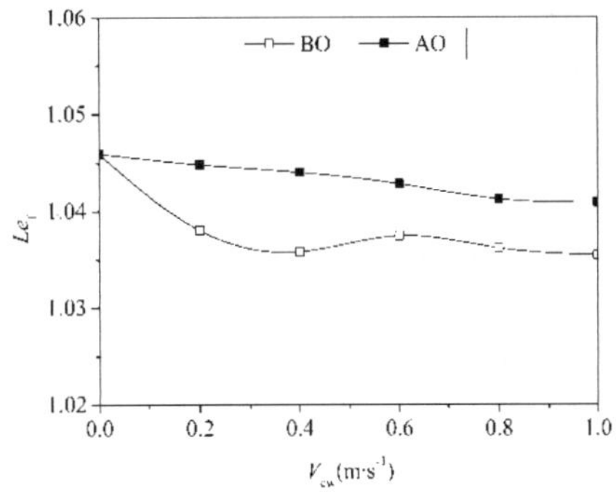

(d)

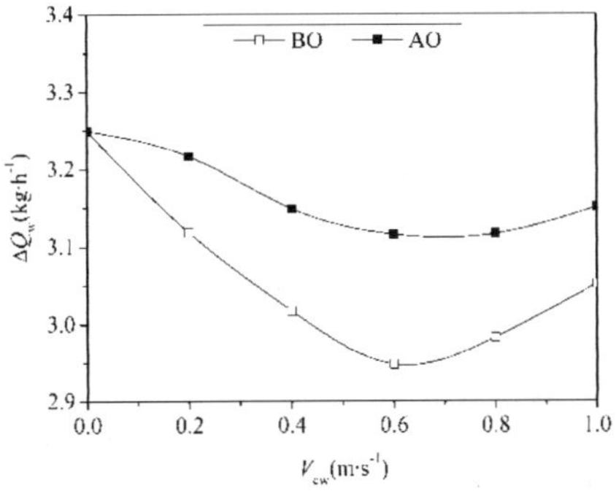

(d)

Figure 4. Crosswinds effect on NDWCT performance before and after inlet airflow optimization. BO and AO represent the performance before and after inlet airflow optimization respectively. (a) cooling efficiency, (b) airflow uniformity coefficient, (c) airflow rate, (d) Lewis factor, (e) evaporation loss

4.2. Airflow rate and airflow uniformity coefficient

Airflow rate G is an index of cooling performance and bigger G represents better performance. According to the balance of heat transfer between air and water, G has the equation of

$$G = \frac{c_w \rho_w Q(T_{wi} - T_{wo})}{K \rho_a (i_{ao} - i_{ai})} \tag{5}$$

where G has the unit of m³/s, c_w is the specific heat of water, kJ/(kg·℃). ρ_w is water density, kg/m³. Q is flow rate of circulating water. ρ_a is air density. i_{ai}、i_{ao} is specific enthalpy of air in and out of tower respectively, kJ/kg. K is the evaporation coefficient that is decided by temperature of outlet water.

In order to analyze the uniformity of circumferential inlet airflow quantitatively, a new parameter of C_u named airflow uniformity coefficient is introduced to with the formula as follows:

$$C_u = \frac{\frac{1}{n}\sum_{i=1}^{n} v_{a,i}}{\frac{1}{n}\sum_{i=1}^{n} v_{a,i} + \sqrt{\frac{1}{n}\sum_{i=1}^{n}\left(v_{a,i} - \frac{1}{n}\sum_{i=1}^{n} v_{a,i}\right)^2}} \tag{6}$$

In Eq. (6), n is number of test points that is taken as 8 here. $v_{a,i}$ is the velocity of inlet air at the test point of i, m/s. C_u related to average velocity of inlet airflow and standard error demonstrates the degree of deviation of inlet airflow.

As the definition of Eq. (6), the range of C_u is 0~1. In theory, if there is not crosswind, inlet airflow is totally even circumferentially i.e. $v_{a,I}$ is equal everywhere, with C_u=1. But existence of crosswind disturbs the uniformity of inlet airflow, making C_u<1 as a result. If crosswind gets extremely large that results in a great difference of inlet air velocities around tower, the C_u can be almost 0. At the condition of certain airflow rate, bigger C_u represents more uniform heat and mass transfer and better cooling performance, in opposite, smaller C_u shows bigger difference of heat and mass transfer and more points of weak heat exchange that deteriorates the global cooling efficiency.

Before the optimization, it can be seen in Fig. 5(a), 4(b) and 4(c) that the velocities of circumferential inlet airflow are almost the same without crosswind, at the condition of that, C_u =0.97 and G reaches the maximum of 165.27 m^3/h. While crosswind increases inlet airflow of windward and reduces that of leeward sharply, making circumferential inlet airflow deviates from the no crosswind condition and totally reducing C_u and G. As v_{cw} up to 0.6 m/s, outflow is observed at leeward, C_u reduces to 0.44, and G reaches the minimum of 154.64 m^3/h, which is 6.43% lower than the maximum. When v_{cw} exceeds 0.6 m/s, the uniformity of inlet airflow gets worse, nevertheless airflow rate presents of recovery trend, e.g. C_u = 0.3, G = 160.45 m^3/h, at the point of v_{cw} = 1.0 m/s.

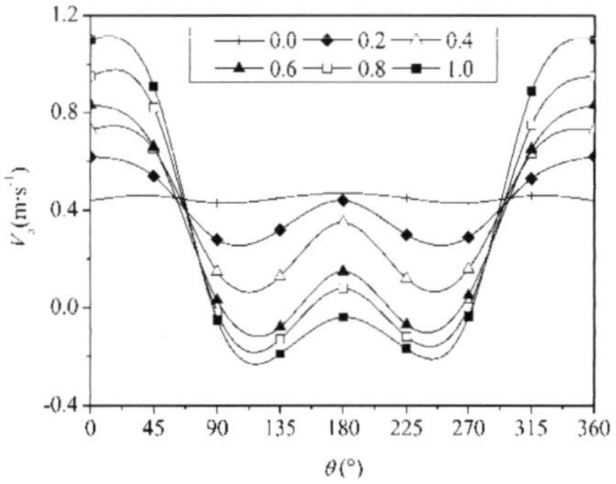

(a)

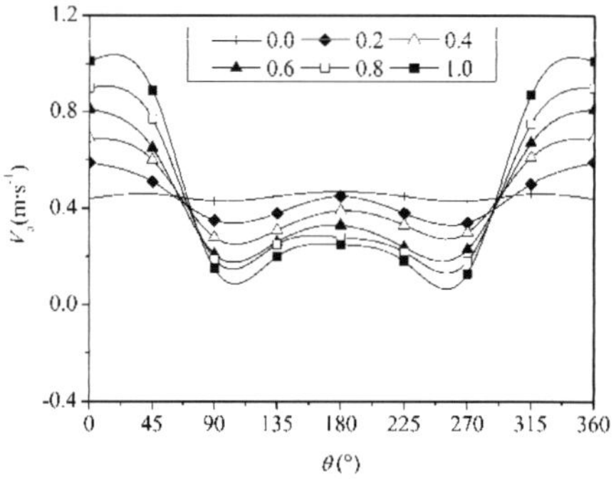

(b)

Figure 5. Crosswinds effect on circumferential distribution of the NDWCT inlet air velocity. (a) before inlet airflow optimization, (b) after inlet airflow optimization

After installing air deflectors at tower inlet, as showed in Fig. 5(b), 4(b) and 4(c), the general changing trend is the same as that of before optimization, but the deviation of inlet airflow of windward and leeward is obviously lower than that of no crosswind condition. However, outflow is not observed at the studied range of crosswind velocities and airflow uniformity coefficient increases universally, airflow rate has a relatively big rise as well. When v_{cw} = 0.6 m/s, C_u = 0.65, G = 160.43 m^3/h.

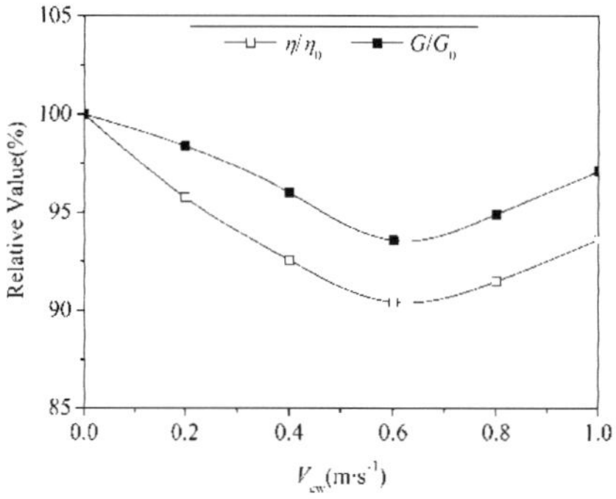

Figure 6. Comparison between the relative variation of η and G under crosswind conditions

4.3. *Relative variation of η and G*

Fig. 6 shows the comparison between the relative variation of η and G under crosswind conditions, in which, η_0 and G_0 of no crosswind case are used as benchmarks. It can be seen that the change of η / η_0 is bigger than G / G_0, e.g. when v_{cw} = 0.6 m/s, η decreases 9.57%, while G only lessens 6.43%, which indicates that the reduction of G is not the only reason of the drop of performance but also attributed to the circumferential nonuniformity of inlet airflow that discomfit the heat and mass transfer in tower. Fig. 7 compares the relative variation of η and G before and after inlet airflow optimization. From an opposite sight ,because of the optimization, increase of η is also bigger than G, e.g. when $v_{cw} = 0.6$ m/s, η increases 5.88%, while G only raises 3.74%. This means that the air deflectors not only increase airflow rate but also enhance the uniformity of inlet air and equalize heat and mass transfer. So the performance change of cooling tower under crosswind is the result of variation of G and uniformity of inlet air.

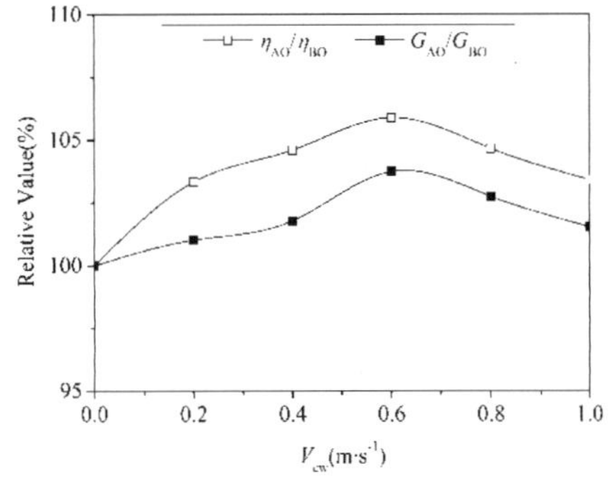

Figure 7. Comparison between the relative variation of η and G before and after inlet airflow optimization

4.4. *Lewis factor and evaporation loss*

Lewis factor Le_f represents the relationship of intensity of heat and mass transfer in the process of evaporation with the formula,

$$Le_f = \frac{k_h}{c_p k_m} \tag{7}$$

where c_p is constant-pressure specific heat of moisture air, kJ/(kg·℃); k_h is heat transfer coefficient, kW/(m^2·℃); k_m is mass transfer coefficient, kg/(m^2·s)。 It can be seen in equation (7) that bigger Le_f indicates more intense heat transfer while smaller Le_f implies faster mass transfer.

In the wet cooling towers, evaporation cooling is dominant and the loss of water ΔQ_w can be calculated by:

$$\Delta Q_w = G(\chi_{ao} - \chi_{ai}) \tag{8}$$

where χ_{ai} and χ_{ao} represent moisture content of air going in and out of tower respectively, kg/kg.

In the paper of [23], Le_f ranges from 0.5 to 1.3, taking 1.0 as an approximate value. Le_f of Fig.4(d) ranges 1.03~1.05, which is dovetailed nicely with [23].Additionally, it can be seen in Fig.4(d) that Le_f enlarges after optimization which means the air deflectors relatively have more function in improving heat transfer and minifies the ratio of latent heat in the total heat load, while water loss caused by evaporation still increases. As in Fig.4 (e), owing to the optimization, water loss increases 5.69% when $v_{cw} = 0.6$ m/s, which is the cost of improving cooling efficiency.

5. Conclusion

1) A new approach of installing air deflectors around tower inlet circumferentially is proposed to optimize the inlet airflow of NDWCT. Then the effect of air deflectors on tower performance under crosswind conditions is investigated through a hot model test. It is shown that crosswind has a great influence on cooling tower performance before optimization, cooling efficiency decreases firstly and then increases as the crosswind gets bigger. While after optimization, cooling efficiency raises intensely at all crosswind conditions and the change with wind gets smoothly. So installing air deflectors could ameliorate the adverse effect of crosswind on cooling performance.

2) In order to analyze the uniformity of circumferential inlet airflow quantitatively, inlet airflow uniformity coefficient C_u is defined. The model test result shows that without crosswind, inlet air is very uniform and C_u=0.97, air flow rate is biggest as well. Existence of crosswind deviate the inlet air, and leeside inlet flow decreases, even change into out flow, both C_u and G get lower sharply. Installing air deflectors makes inlet flow of circumference more uniform and increases C_u and G.
3) Analysis of relative change of cooling efficiency and air flow rate before and after optimization indicates two aspects of crosswind effect, firstly, crosswind reduces airflow rate, and secondly, crosswind destroys the uniformity of circumferential inlet airflow so as to make heat transfer of inner tower uneven and deteriorates cooling performance as a result. Air deflectors optimize both of the two aspects and improve cooling efficiency.
4) As Le_f increases after the optimization, air deflectors have more function on enhancing heat transfer than mass transfer and reduce the ratio of evaporation cooling; however the total water loss still gets bigger.

6. References

[1] Li Xiuyun, Yan Junjie,Lin Wanchao. Study on thermo-economics diagnosis method and index evaluation system for the cold-end system in steam power unit[J]. Proceedings of the CSEE, 2001, 21(9): 94-99.
[2] Kloppers J C, Kröger D G. A critical investigation into the heat and mass transfer analysis of counterflow wet-cooling towers[J]. International Journal of Heat and Mass Transfer, 2005, 48(3-4): 765-777.
[3] Fisenko S P, Petruchik A I, Solodukhin A D. Evaporative cooling of water in a natural draft cooling tower[J]. International Journal of Heat and Mass Transfer, 2002, 45(23): 4683-4694.
[4] Hawlader M N A, Liu B M. Numerical study of the thermalhydraulic performance of evaporative natural draft cooling towers[J]. Applied Thermal Engineering, 2002, 22(1): 41-59.
[5] Williamson N, Armfield S, Behnia M. Numerical simulation of flow in a natural draft wet cooling tower-the effect of radial thermofluid fields[J]. Applied Thermal Engineering, 2008, 28(2-3): 178-189.
[6] Smrekar J, Oman J, Sirok B. Improving the efficiency of natural draft cooling towers[J]. Energy Conversion and Management, 2006, 47(9-10): 1086-1100.
[7] Muangnoi T, Asvapoositkul W, Wongwises S. An exergy analysis on the performance of a counterflow wet cooling tower[J]. Applied Thermal Engineering, 2007, 27(5-6): 910-917.
[8] Liu Zhiyun, Wang Dong, Lin Zonghu. Numerical analysis of the influence of side wind on the performance of the direct air cooling tower with natural ventilation[J]. Journal of Power Engineering, 2008, 28(6): 915-919.
[9] Zhang Xiaodong, Wang Qingzhao. Impact of side wind on the cooling performance of natural draft air cooling tower [J]. Electric Power, 1999, 32(6): 34-36, 47.
[10] Zhao Zhenguo, Shi Jinling, Wei Qingding, et al. The engineering improvement for weakening the bad effect of natural wind on the dry cooling towers[J]. Journal of Applied Sciences, 1998, 16(1): 112-120.
[11] Derksen D D, Bender T J, Bergstrom D J, et al. Study on the effects of wind on the air intake flow rate of a cooling tower: Part 1. Wind tunnel study[J]. Journal of Wind Engineering and Industrial Aerodynamics, 1996, 64(1): 47-59.
[12] Bender T J, Bergstrom D J, Rezkallah K S. Study on the effects of wind on the air intake flow rate of a cooling tower: Part 2. Wind tunnel study[J]. Journal of Wind Engineering and Industrial Aerodynamics, 1996, 64(1): 61-72.
[13] Bender T J, Bergstrom D J, Rezkallah K S. Study on the effects of wind on the air intake flow rate of a cooling tower: Part 3. Numerical study[J]. Journal of Wind Engineering and Industrial Aerodynamics, 1996, 64(1): 73-88.
[14] Zhai Z, Fu S. Improving cooling efficiency of dry-cooling towers under cross-wind conditions by using wind-break methods[J]. Applied Thermal Engineering, 2006, 26(10): 1008-1017.

[15] Al-Waked R, Behnia M. CFD simulation of wet cooling towers[J]. Applied Thermal Engineering, 2006, 26(4): 382-395.
[16] Al-Waked R, Behnia M. Enhancing performance of wet cooling towers[J]. Energy Conversion and Management, 2007, 48(10): 2638-2648.
[17] Gao M, Sun F Z, Wang K, et al. Experimental research of heat transfer performance on natural draft counter flow wet cooling tower under cross-wind conditions[J]. International Journal of Thermal Sciences, 2008, 47(7): 935-941.
[18] Zhao Yuanbin, Sun Fengzhong, Wang Kai, et al. Numerical analysis of crosswind effect on wet cooling tower aerodynamic field[J]. Nuclear Power Engineering, 2008, 29(6): 35-40.
[19] Zhao Yuanbin, Sun Fengzhong, Wang Kai, et al. Numerical analysis of the cooling performance of wet cooling tower with cross wall[J]. Proceedings of the CSEE, 2009, 29(8): 6-13.
[20] Zhou Lanxin, Jiang Bo, Chen Sumin. Numerical simulation for thermal performance of natural draft wet cooling tower[J]. Journal of Hydraulic Engineering, 2009, 40(2): 208-213.
[21] Jiang Bo, Zhou Lanxin. Numerical study of installing cross-windbreak on natural draft cooling tower[J]. Turbine Technology, 2010, 52(3): 165-168, 172.
[22] Shi Youji. Cooling tower operation and test[M]. Beijing: China Water-Power Press, 1990: 183-188.
[23] Kloppers J C, Kroger D G. The Lewis factor and its influence on the performance prediction of wet-cooling towers[J]. International Journal of Thermal Sciences, 2005, 44(9): 879-884.

IOP Conf. Series: Earth and Environmental Science **114** (2018) 012006 doi:10.1088/1755-1315/114/1/012006

Application of porous medium for efficiency improvement of a concentrated solar air heating system

Boonrit Prasartkaew

Combustion and Solar Energy Laboratory (CASE Lab.), Faculty of Engineering, Rajamangala University of Technology Thanyaburi, Klong 6, Thanyaburi, Patumthani 12110, Thailand
E-mail: boonrit.p@en.rmutt.ac.th

Abstract. The objective of this study is to evaluate the thermal efficiency of a concentrated solar collector for a high temperature air heating system. The proposed system consists of a 25-m^2 focused multi-flat-mirror solar heliostat equipped with a porous medium solar collector/receiver which was installed on the top of a 3-m tower, called 'tower receiver'. To know how the system efficiency cloud be improved by using porous medium, the proposed system with and without porous medium were tested and the comparative study was performed. The experimental results reveal that, for the proposed system, application of porous medium is promising, the efficiency can be increased about 2 times compared to the conventional one. In addition, due to the porous medium used in this study was the waste material with very low cost. It can be summarized that the substantial efficiency improvement with very low investment cost of the proposed system seem to be a vital measures for addressing the energy issues.

1. Introduction

Currently, energy seem to be one of the most essential human needs to survive, all activities need energy. Literature reveals that the energy consumption is growing rapidly in many countries and almost all the energy resources are fossil based energy which will be depleted by the next few ten years. Moreover, all of the energy conversion processes of fossil based energy release poison emission and/or green-house gas (GHG) to the environment. Relevant to the trend of energy consumption and demand, the energy crisis and environmental problems are now become the serious issues for many countries around the world. One of the best ways to simultaneously overcome these serious issues is the utilization of renewable energy together with energy conservation measures [1, 2]. Therefore, as renewable energy becomes a global priority, solar power is an increasingly viable solution to the world's energy demand and environmental issues. From the global trend with projection forecasting, renewables account for a rising share of the world's total electricity supply, the renewable share of world electricity generation grows from 22% in 2012 to 29% in 2040 [3].

Solar energy is the biggest renewable energy resource of the world with the extra-terrestrial irradiance, radiation falls on a unit area normal to the sun in unit time at the top of the earth's atmosphere, with the accepted amount of 1,376 W/m^2 and it will not run out [4]. The terrestrial solar radiation is different for each locations around the world as the solar insolation map shown in Figure 1. However, due to solar energy can be utilized only during day time, day length is one of the most important parameter for system design and selection. This parameter mainly depends on the latitude and altitude of the site location.

IOP Conf. Series: Earth and Environmental Science **114** (2018) 012006 doi:10.1088/1755-1315/114/1/012006

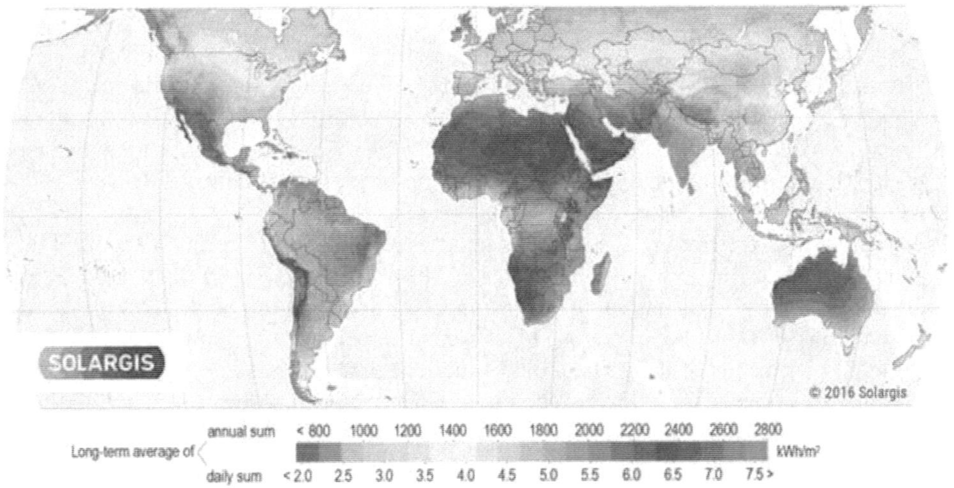

Figure 1. Solar insolation map [6].

Figure 2 shows the amount of solar energy in hours, received each day on an optimally tilted surface during the worst month of the year (based on accumulated worldwide solar insolation data). For every an hour, enough sunlight energy reaches the earth to meet the world's energy demand for a whole year [5]. To reduce the fossil fuel consumption and CO ^{2}emissions, generating sustainable heat/power from solar, a largest resource and readily available to everyone, can be possibly opted as the best way of above problems mitigation.

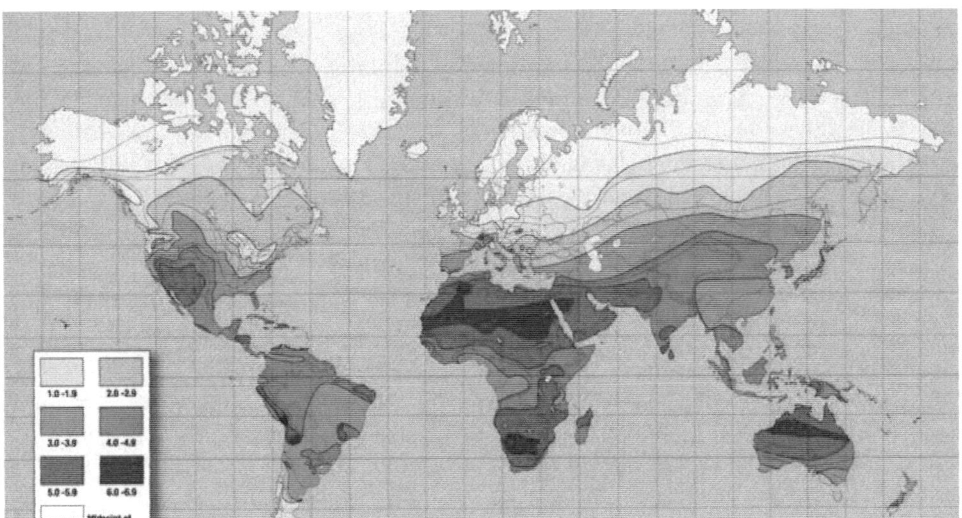

Figure 2. The amount of solar energy in hours [7].

Currently, there are many applications of solar thermal energy, e.g.: domestic and industrial heating and cooling systems, drying, distillation and solar cooking, etc. Solar thermal collector is a device that utilizes the solar radiation to heat a fluid (usually water or air), which can then be used for suitable aforesaid applications. But, when higher output temperatures are required, the concentrated solar collector is taken into consideration. In order to do that, an optical device could be placed between the radiant source and the absorbing surface or receiver. This device that optically reflect and focus incident solar energy onto a smaller receiving area. There are many types of concentrating collector, e.g.: compound parabolic concentrators (CPC), parabolic trough, parabolic dish and central receiver, etc. Seasonably, the more complex of such systems, the more expensive investment is required.

Like flat plate solar collector, the thermal performance of concentrating collectors depend on the useful energy delivered by the collector, this mainly depends on the collector's heat removal factor (F_R) and overall heat loss coefficient (U_L). To improve the efficiency of solar collector, by enhancing the heat removal factor and reducing heat loss, the application of porous medium (PM) was proposed for this current study.

From the literatures review, there are some works regarding the PM solar receiver reported by [8-14]. It can be summarized that they reported the results of theoretical and experiment studies on the PM receiver of the parabolic trough and multi-dish concentrator using liquid as working medium. In addition, the previous work as reported in [15-17] have developed the control system for a solar heliostat positioning by means of the calculation of the real-time sun position to redirect the sun rays for striking it onto the target (called 'tower receiver'). The developed software was implemented with an 11.2-m^2 solar heliostat. The experimental results demonstrated that the reflection accuracy of the proposed focused multi-flat-mirror solar heliostat was acceptable.

Regarding the aforesaid literature reviews and previous works, there are some research gaps can be contributed from this study: nobody has reported about the performance of the solar collector using PM receiver for air heating system, re-using of metal waste from lathing process (metal swarf) as the porous medium, etc. Moreover, the different of between the receiver of parabolic-trough/multi-dish-concentrator and tower receiver are that the former one was fixed together with the reflector while the later one is on ground supported tower and can carry heavier load. Therefore, the objective of this study was to evaluate the thermal efficiency of a focused multi-flat-mirror solar heliostat equipped with a PM tower receiver for a high temperature air heating system was proposed. To know how the system efficiency cloud be improved by using porous medium, the proposed system with and without porous medium were tested and the comparative study was performed.

2. Experimental setup

The experimental system, as the schematic diagram shown in Figure 3, was setup at latitude of 14.03N and longitude of 100.72E. This system comprises of two main parts, a PM tower receiver for a high temperature air heating system and a focused multi-flat-mirror solar heliostat as shown in Figure 4 and 5, respectively. The centre of receiver and heliostat are on the same level at 3 m from the ground level. The receiver was located in the south of heliostat with the distance of 20 m.

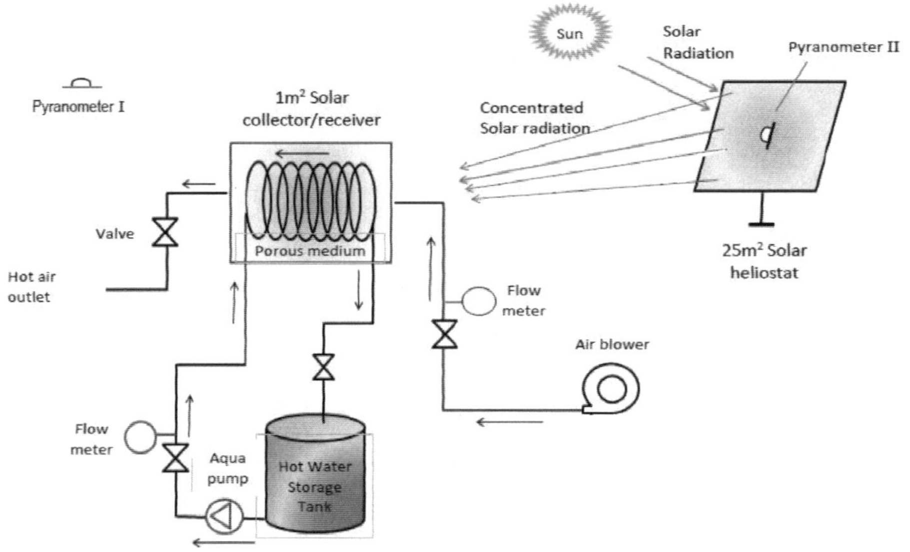

Figure 3. Experimental Setup.

IOP Conf. Series: Earth and Environmental Science **114** (2018) 012006 doi:10.1088/1755-1315/114/1/012006

A $0.7x0.7x0.7m^3$ insulated solar tower receiver, as shown in Figure 4 was made of metal sheet. A single glass cover was used to reduce the losses of the solar collector. Air was supplied to the receiver via an adjustable flow rate air blower. The porous medium (as shown in Figure 4 (c) was fully filled in the receiver box.

Figure 4. The tower receiver (a) back side, (b) front view and (c) porous medium (metal swarf).

The $5x5m^2$ focused multi-flat-mirror solar heliostat, as shown in Figure 5, was fabricated by using low cost material available in the market. Its working principle is the same as the system presented by [15-17] but its effective aperture area is bigger. A solar radiation redirection system with a heliostat, 289 flat fixed mirrors, installed on a tower, was proposed. Each $0.3x0.3m^2$ flat mirror are attached on an individual adjustable mounting set, so the desired tilting angle achieves more flexibility. The focus of the heliostat was adjusted under the real solar radiation during 1 to 2 hours before and after solar noon.

Figure 5. The focused multi-flat-mirror solar heliostat used in this study.

3. Performance evaluation

For evaluating the performance of the proposed system, the experiments were done during 9am to 4pm. The global solar radiation incident on the horizontal and on the heliostat plan were measured using two pyranometers (as shown in Figure 6 (a)). The temperatures of working fluid were measured using K-type thermocouples. The experimental data were collected via a data logger, Graphtec model GL820 (as shown in Figure 6 (b)), with the recording interval of 1 minute.

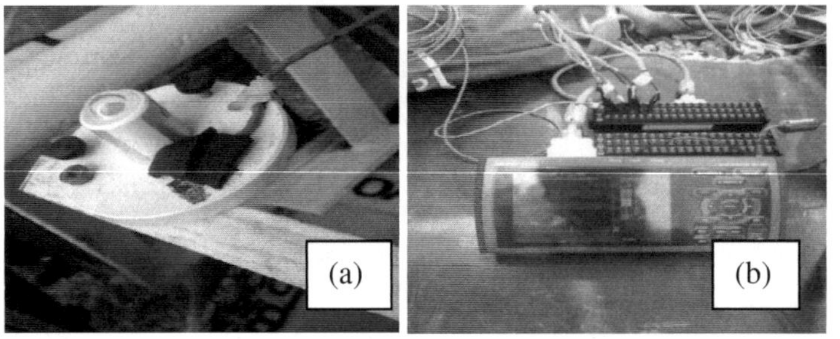

Figure 6. (a) Pyranometer and (b) data logger used in this study.

The ISO 9459-2 [18] was conducted for the performance test of the proposed system. To know the system efficiency, the ratio of useful heat output rate ($\dot{Q}_u$) to the total input solar energy ($A_c G_T$) was calculated as the following equations [19]:

$$\dot{Q}_u = \rho \dot{V} c_p (T_{c,o} - T_{c,i}) \quad (1)$$

and

$$\eta_c = (\dot{Q}_u / A_c G_T) \times 100\% \quad (2)$$

where,

$\dot{Q}_u$	=	useful heat output rate [W]
ρ	=	air density [= 1.127 kg/s]
$\dot{V}$	=	air volume flow rate [m^3/s]
c_p	=	specific heat of air [= 1.007 kJ /kg.K]
$T_{c,o}$	=	outlet temperature of tower receiver [°C]
$T_{c,i}$	=	inlet temperature of tower receiver r [°C]
η_c	=	tower receiver efficiency [-]
A_c	=	effective aperture area of solar heliostat [m^2]
G_T	=	solar irradiance incident on heliostat plane [W/m^2]

4. Experimental results and discussions

The experiments were done for more than ten days in January to February 2017, every other day for with- and without-porous medium. The air volume flow rate was fixed at 0.04 m^3/s, this value was controled by manually adjusting the inlet air damper actuator of the blower every half an hour.

4.1. The tower receiver without-porous medium

Figure 7 shows the experimental results when the tower receiver without porous medium was conducted. Figure 7 (a) shows the total solar irradiance and ambient temperature of the clear sky experimental day on January 29, 2017. The data show that the daily average irradiance and ambient temperature were 930 W/m^2 and 32 °C, respectively. Figure 7 (b) shows the air temperatures at the inlet and outlet and the glass cover of the tower receiver, respectively. The calculated mass flow rate of air was 0.05 kg/s. The average hot air temperature of this day was about 99°C. Figure 7 (c) shows the efficiency of the air heating system without porous medium application, calculated as equation (2). The experimental results demonstrated that the average thermal efficiency of about 12% was obtained.

4.2. The tower receiver with-porous medium

Figure 8 shows the experimental results when the tower receiver without porous medium was conducted. Figure 8 (a) shows the total solar irradiance and ambient temperature of the clear sky experimental day on February 12, 2017. The data show that the daily average irradiance and ambient temperature were 950 W/m^2 and 32 °C, respectively. Figure 8 (b) shows the air temperatures at the inlet and outlet and the glass cover of the tower receiver, respectively. With the same air supplied mass flow rate of 0.05 kg/s. The average hot air temperature of this day was about 122°C. Figure 8 (c) shows the efficiency of the air heating system with porous medium application. The experimental results demonstrated that the average thermal efficiency of about 22% was obtained.

From the experimental results, it was found that the performance of the proposed system can be improved significantly by using the porous medium. At the constant air supply mass flow rate of 0.05 kg/s, the thermal efficiency was increased from 99°C to 122°C (increased by 23.23%).The system efficiency can be increased from 12% to 22% (increased by 83.33%). As the information that porous media is a material containing pores typically filled with a fluid (air). Thus, it causes increase in heat transfer in the heat extraction process between solar thermal and air. It exhibit enhanced thermal heat transfer process, such as higher thermal intensity, conductivity, convective heat transfer coefficients and emissivity compared to the conventional single phase process.

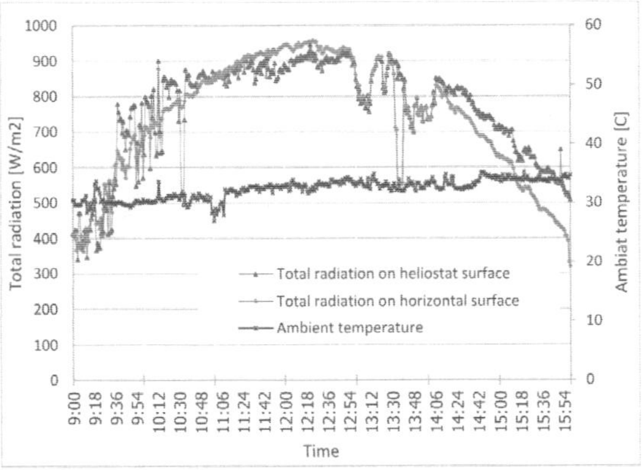

(a)

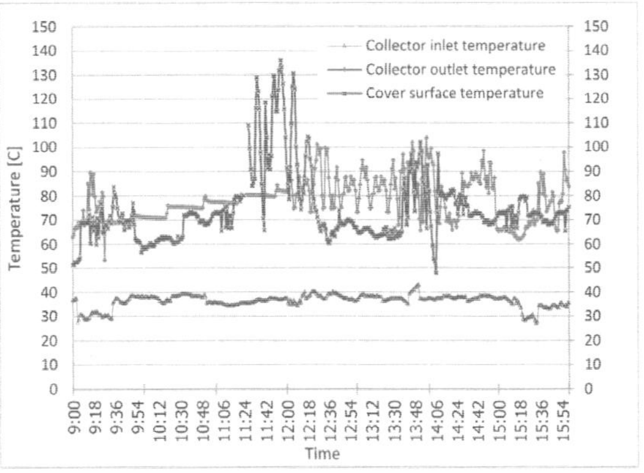

(b)

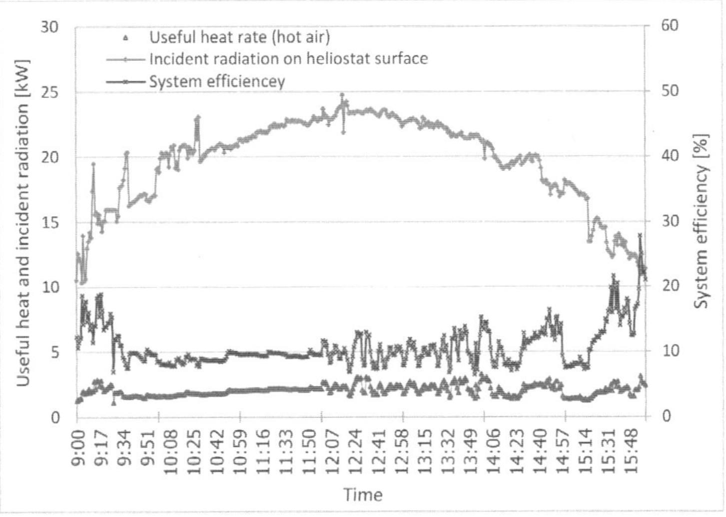

(c)

Figure 7. The experimental results when the tower receiver without porous medium was conducted
(a) total solar irradiance and ambient temperature of the clear sky experimental day
(b) air temperatures at the inlet and outlet and the glass cover of the tower receiver
(c) efficiency of the air heating system without porous medium application.

IOP Conf. Series: Earth and Environmental Science **114** (2018) 012006 doi:10.1088/1755-1315/114/1/012006

5. Conclusions

Experimental study was conducted to prove the hypothesis that porous medium can improve the performance of a concentrated solar collector for a high temperature air heating system. Hot air temperature and system thermal efficiency were compared between with- and without-porous medium tower receivers at constant controlled variables. The experimental results reveal that, for the proposed system, application of porous medium is promising, the efficiency can be increased about 2 times compared to the conventional one. In addition, due to the porous medium used in this study was the waste material with very low cost. It can be summarized that the substantial efficiency improvement with very low investment cost of the proposed system seem to be a vital measures for addressing the energy issues.

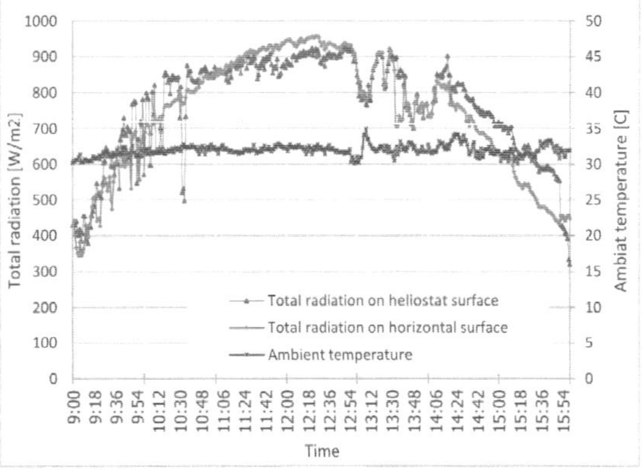

(a)

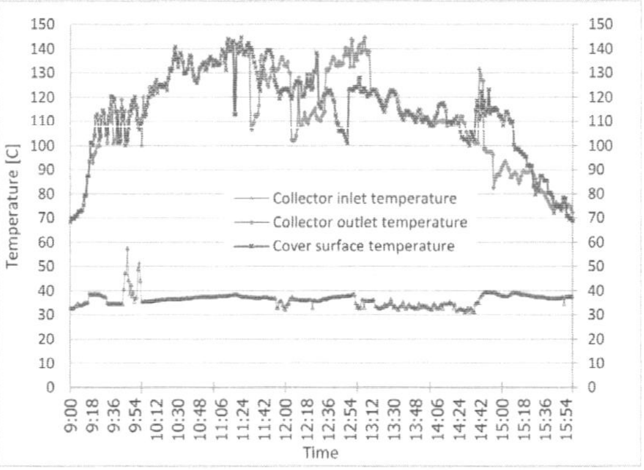

(b)

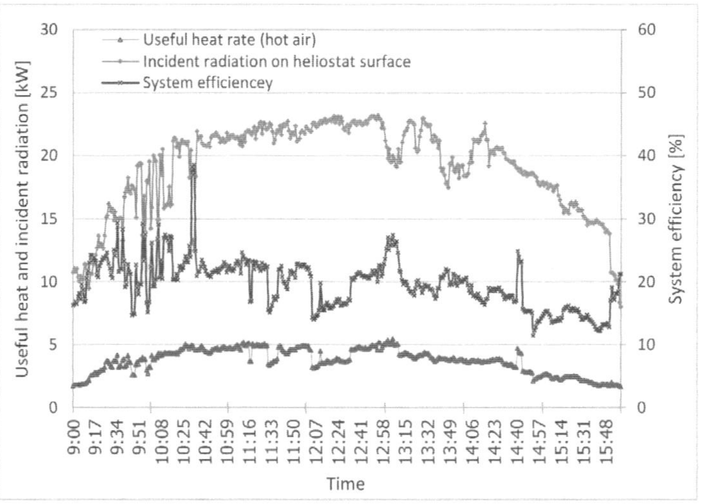

(c)

Figure 8. The experimental results when the tower receiver with porous medium was conducted (a) total solar irradiance and ambient temperature of the clear sky experimental day (b) air temperatures at the inlet and outlet and the glass cover of the tower receiver (c) efficiency of the air heating system with porous medium application.

6. References

[1] B. Prasartkaew, S. Kumar, A low carbon cooling system using renewable energy resources and technologies, *Energy and Buildings* 42 (2010) , pp. 1453–1462

[2] B. Prasartkaew, S. Kumar, Experimental Study on the Performance of a Solar-Biomass Hybrid Air-Conditioning System, *Renewable Energy* 57 (2013), pp. 86-93

[3] U.S. Energy Information Administration (EIA). *International Energy Outlook 2016*. Available online: (www.eia.gov/ieo). [Download: July 15, 2017]

[4] Julie Richards, Solar energy, *Cavendish Square Publishing*, (2009) [online]. Available: http:// http://cdn-media.macmillan.com.au/mea/downloadpdfs/9781420274219.pdf, [Download: July 15, 2017]

[5] Xiudong Wei, Zhenwu Lu, Zhifeng Wang, Weixing Yu, Hongxing Zhang, Zhihao Yao, A New Method for the Design of the Heliostat Field Layout for Solar Tower Plant, *Renewable Energy* 35(2010) 1970-1975

[6] Solargis. *World solar resource maps*. [online]. Available: http://solargis.com/products/maps-and-gis-data/free/download/world, [Download: July 15, 2017]

[7] Alternative Energy Store Inc. *World Solar Insolation Values*. [online]. Available: https://www.altestore.com/howto/solar-insolation-map-world-a43/, [Download: July 15, 2017]

[8] Wang P, Liu DY, Xu C. Numerical study of heat transfer enhancement in the receiver tube of direct steam generation with parabolic trough by inserting metal foams. *Appl Energy* 2013; 102, 49–60

[9] Wang FQ, Tan JY, Wang ZQ. Heat transfer analysis of porous media receiver with different transport and thermos-physical models using mixture as feeding gas. *Energy Convers Manag* 2014; 83, pp.59–66

[10] Wang FQ, Tan JY, Shuai Y, Tan HP, Leng Y. Thermal performance analysis of porous medium solar receiver with quartz window to minimize heat flux gradient. *Solar Energy* 2014; 108, pp.48–59

[11] Wang FQ, Tan JY, Shuai Y, Tan HP, Chu SX. Thermal performance analyses of porous media solar receiver with different irradiative transfer models. *Int J HeatMass Transf* 2014; 78, pp.7–16

[12] Reddy KS, Ravi KK, Ajay CS. Experimental investigation of porous disc enhanced receiver for solar parabolic trough collector. Renew Energy 2015; 77, 8–19

[13] Chen X, Xia XL, Meng XL, Dong XH. Thermal performance analysis on a volumetric solar receiver with double–layer ceramic foam. *Energy Convers Manag* 2015; 97, pp.2–9

[14] Wang FQ, Guan ZN, Tan JY, Ma LX, Yan ZY, Tan HP. Transient thermal performance response characteristics of porous–medium receiver heated by multi–dish concentrator. *Int Commun Heat Mass* 2016; 75, pp.36–41

[15] Boonrit Prasartkaew, Amnaj Ngernplabpla and Tinnakorn Poowadin, Control System of a Concentrated Solar Heliostat, *RMUTI Journal Special Issue 1 2015, The 6th International Conference on Science, Technology and Innovation for Sustainable Well-Being (STISWB VI)*, (2015), pp. 55-62

[16] Amnaj Ngernplabpla and Boonrit Prasartkaew, Experimental Investigation on the Intensity Efficiency of a Solar Heliostat System, *2013 Alternative Energy in Developing Countries and Emerging Economies (AEDCEE)*, 30-31 May 2013, Pullman Bangkok King Power, Bangkok, Thailand

[17] Boonrit Prasartkaew and Amnaj Ngernplabpla, Investigation on the Performance of a Paraboloids Heliostat for Concentrated Central Receiver Solar Collector, *2013 International Conference on Frontiers of Environment, Energy and Bioscience (ICFEEB 2013)*, Beijing, China, October 24-25, 2013

[18] ISO. (1995). Domestic water heating systemsPart 2, *outdoor test methods for system performance characterization and yearly performance prediction of solar-only systems. ISO 9459-2*

[19] Duffie, J.A. and Beckman, W.A. (1991). *Solar Engineering of Thermal Processes*, second ed. John Wiley, New York

IOP Conf. Series: Earth and Environmental Science **114** (2018) 012007 doi:10.1088/1755-1315/114/1/012007

Crosswinds Effect on the Thermal Performance of Wet Cooling Towers Under Variable Operating Conditions

You Liang Chen[1], Yong Feng Shi[1], Jian Gang Hao[1], Hao Chang[1], Feng Zhong Sun[2]

[1]Huadian Electric Power Research Institute, No. 2 Xiyuan Ninth Road, Hangzhou 310030, China
[2]Shandong University, No. 17923 Jingshi Road, Jinan 250061, China

E-mail: cyl_sdu@163.com

Abstract. In order to quantitatively analyze the influence of the variable operating parameters on the cooling performance of natural draft wet cooling towers (NDWCTs), a hot model test system was set up with adjustable ambient temperature and humidity, circulating water flowrate and temperature. In order to apply the hot model test results to the real tower, the crosswind Froude number is defined. The results show that the crosswind has a negative effect on the thermal performance of the cooling tower, and there is a critical crosswind velocity corresponding to the lowest cooling efficiency. According to the crosswind Froude number similarity, when the ambient temperature decreases, or the circulating water flowrate and temperature increase, the cooling tower draft force will increase, and the critical crosswind velocity will increase correspondingly.

1. Introduction

Natural draft wet cooling towers are widely used to cool the circulating water in thermal power plants. Existing studies have shown that the environmental crosswind could reduce the cooling tower performance, and can even cause the unit trip in extreme cases. Russell [1] pointed out that the crosswind can greatly affect the cooling tower performance through the dry tower and industrial cooling tower research. Grange et al. [2] performed prototype and model tower tests on natural draft cross-flow wet cooling towers. The results show that crosswinds can adversely affect cooling performance, and crosswinds will become favorable when the crosswind velocity exceeds a certain value. Bender et al. [3] tested the model tower in the wind tunnel. The results show that crosswinds could seriously affect the air flow in the tower, and would lead to an increment by 45% in the windward side. Al-Waked et al. [4] found that when the crosswind velocity was 7.5 m/s, the tower outlet water temperature was increased by 1.7 K. And the main reason was nonuniform distribution of the air flow in the tower caused by crosswinds. Gao et al. [5] found that the crosswind would adversely affect the cooling water temperature difference and the cooling efficiency by hot model tests, both of which could be reduced by a maximum of 6% and 5%. Zhao et al. [6] established a three-dimension numerical model of the cooling tower. It was found that the crosswind increased the lateral air flowrate and enhanced the cooling performance of the rain zone, but the longitudinal air flowrate was reduced, which was not conducive to heat and mass transfer in the fill zone, and finally reduced overall performance of the cooling tower. Zhou et al. [7] performed a numerical analysis on wet cooling towers. The results show that the crosswind has a negative effect on the water temperature of the tower, and would lead to a maximum water temperature at the crosswind velocity of 6 m/s, which is 1.34 ℃ higher than that without crosswinds.

In summary, crosswinds effect on the thermal performance of wet cooling towers were studied by means of hot model tests and numerical methods respectively [5, 6]. However, the former literature [5] only discussed the adverse effect of the crosswind on the cooling tower performance, and did not further explore the changing law of the crosswinds effect under various variable conditions. The latter literature [6] only uses the numerical calculation to study the influence of the varying circulating water parameters. In view of this, this paper intends to study the influence and mechanism of the crosswind on the thermal performance of the wet cooling tower when the parameters such as ambient temperature, circulating water flowrate and tower inlet water temperature are changed through the hot model test, so as to guide the actual operation of the cooling tower.

2. Model test

The model cooling tower is made of plexiglass, and the schematic diagram of the entire test system is shown in Figure 1. When the model test begins, the circulating water is heated to a given temperature in the lower tank, pumped to the upper tank by the circulating pump, and then it flows into the tower. Finally, it falls into the water basin after being cooled, and flows back into the lower tank, to complete a cycle. The detailed description of the hot model test system can refer to the previous literature [8].

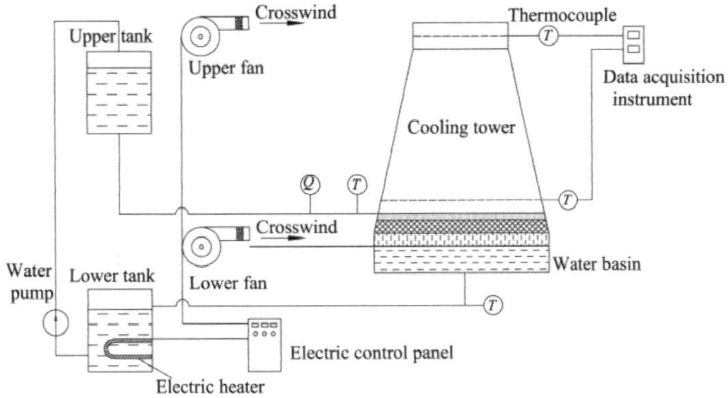

Figure 1. Schematic diagram of the hot model test system

During the test, the ambient temperature is selected as 16, 20 and 24 ℃, the tower inlet water temperature is selected as 40, 50 and 60 ℃, the circulating water flowrate is selected as 4, 6 and 8 L/min, and the crosswind velocity is selected as 0.0, 0.2, 0.3, 0.4, 0.5, 0.6, 0.7 and 0.8 m/s.

In this test, the main parameters to be measured include atmospheric pressure, dry and wet bulb temperature, crosswind velocity, circulating water flow rate, water and air temperature at the inlet and outlet of the tower. The parameters and measuring instruments are detailed in literature [8].

3. Results and discussion

3.1. Cooling efficiency

Cooling efficiency η is often used to indicate the thermal performance of NDWCTs. η is defined as $\eta=(t_1-t_2)/(t_1-\tau)$, where t_1 and t_2 are the tower inlet and outlet water temperature respectively, τ is the wet bulb temperature of the inlet air.

3.2. Crosswinds effect on the cooling efficiency under different ambient temperatures

This section discusses three conditions with the ambient temperature θ_1=16 ℃, 20 ℃ and 24 ℃, the circulating water flowrate Q and the tower inlet water temperature t_1 are set as 6 L/min and 50 ℃ respectively, and the relative humidity φ is maintained at 58 %.

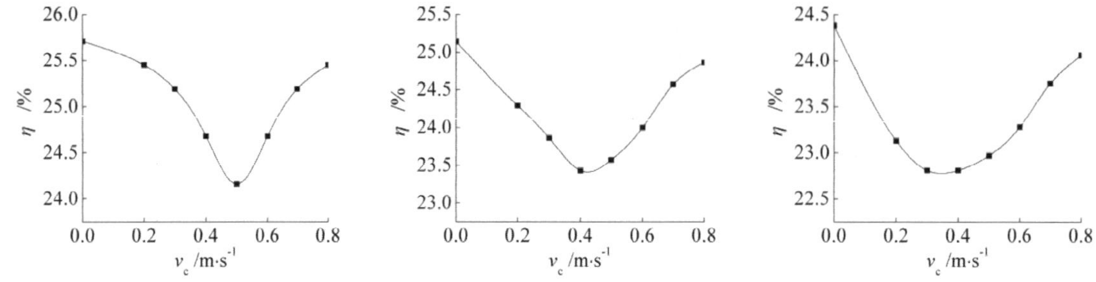

(a) θ_1=16℃ **(b) θ_1=20℃** **(c) θ_1=24℃**

Figure 2. Crosswinds effect on cooling efficiency under different ambient temperatures

Figure 2 shows crosswinds effect on the cooling efficiency at different ambient temperatures. When the crosswind velocity v_c=0 m/s, η=25.7%, 25.1% and 24.4% under three conditions with θ_1=16℃, 20℃ and 24℃. It can be seen that the higher θ_1 is, the lower η will become. This is because when the ambient temperature increases, the temperature difference between the air in the tower and the circulating water will decrease, the contact heat dissipation will decrease and finally the cooling performance will reduce. Therefore, the summer temperature is high, the cooling tower performance is poor, and the winter temperature is low, and is favorable to the cooling performance.

Figure 2 also shows that the trend of crosswinds effect on cooling efficiency is basically the same at different ambient temperatures. In the absence of crosswind, η is higher. When the crosswind increases, η will decrease first and then increase, and there exists the minimum value of η.

Based on the above, a new concept is given to describe this characteristic, that is, for a certain type of cooling tower at a certain structural parameters, environmental parameters and circulating water parameters, v_c corresponding to the minimum η is defined as critical crosswind velocity v_{cr}. v_{cr} corresponds to the minimum η under crosswind conditions, that is, when v_c is lower than v_{cr}, η will gradually reduce, and when v_c is higher than v_{cr}, η will gradually improve.

Figure 2 also reflects the effect of the ambient temperature on the critical crosswind velocity. It can be seen that when θ_1 is 16 ℃ and 20 ℃, v_{cr} is 0.5 m/s and 0.4 m/s, while the corresponding v_{cr} at θ_1=24 ℃ is between 0.3 m/s and 0.4 m/s. This shows that the higher θ_1 is, the smaller v_{cr} will become, and the cooling tower is more susceptible to the adverse effect of environmental crosswinds, that is, the anti-crosswind capacity of cooling tower in the winter is stronger than that in the summer.

3.3. Crosswinds effect on the cooling efficiency under different circulating water flowrates

This section discusses three conditions with Q=4 L/min, 6 L/min and 8 L/min. And t_1=50 ℃, θ_1=20 ℃ and φ=58 %.

Figure 3 shows crosswinds effect on η at different circulating water flowrates. When v_c=0 m/s, η=32.0%, 25.1% and 21.4% under three conditions with Q=4 L/min, 6 L/min and 8 L/min. Obviously, when the other conditions are constant, the smaller Q is, the higher η will become.

It can be seen that there is a critical crosswind velocity in a variety of variable circulation water conditions. It is worth noting that the critical crosswind velocities corresponding to variable circulating water flowrate conditions are not the same. When Q=4 L/min and 6 L/min, the corresponding v_{cr} are 0.3 m/s and 0.4 m/s. And at the case of Q=8 L/min, the corresponding v_{cr} is between 0.5 m/s and 0.6 m/s. It can be seen that v_{cr} will change along with the change of Q, and the larger Q is, the bigger v_{cr} will become. This shows that in the cooling tower operating process, the greater Q is, the stronger the cooling tower will be to withstand the crosswinds effect, but the circulating pump power consumption will increase correspondingly. Therefore, in the actual operation of the cooling tower, it should deal with the relationship between the two factors, so that the cooling tower can meet the needs of cold end while achieving higher operating performance.

IOP Conf. Series: Earth and Environmental Science **114** (2018) 012007 doi:10.1088/1755-1315/114/1/012007

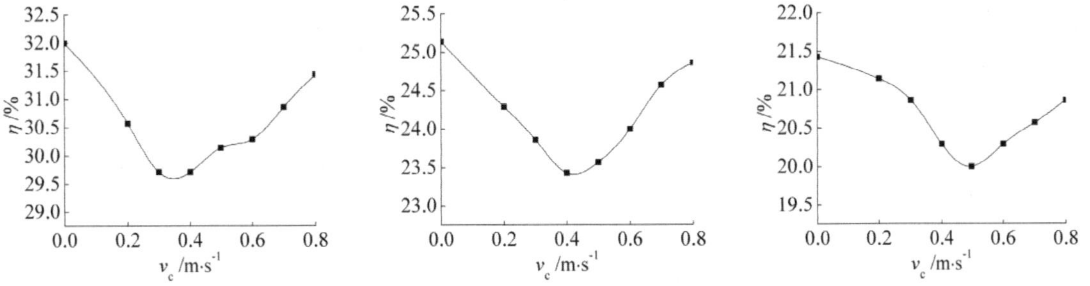

(a) Q=4L/min **(b) Q=6L/min** **(c) Q=8L/min**

Figure 3. Crosswinds effect on cooling efficiency under different circulating water flowrates

3.4. Crosswinds effect on the cooling efficiency under different tower inlet water temperatures

This section discusses three conditions with t_1=40 ℃, 50 ℃ and 60 ℃. And other operating parameters are constant, with Q=6 L/min, θ_1=20 ℃ and φ=58 %.

Figure 4 shows crosswinds effect on the cooling efficiency at different tower inlet water temperatures. When v_c=0 m/s, η=21.6%, 25.1% and 30.7% under the three conditions with t_1=40℃, 50℃ and 60℃. Obviously, when the other parameters are constant, the higher t_1 is, the higher η will become.

It can be seen from Figure 4 that there is a critical crosswind velocity v_{cr} at different tower inlet water temperatures, corresponding to the minimum η under crosswind condition, similar to the variable ambient temperature and circulation water flowrate conditions above.

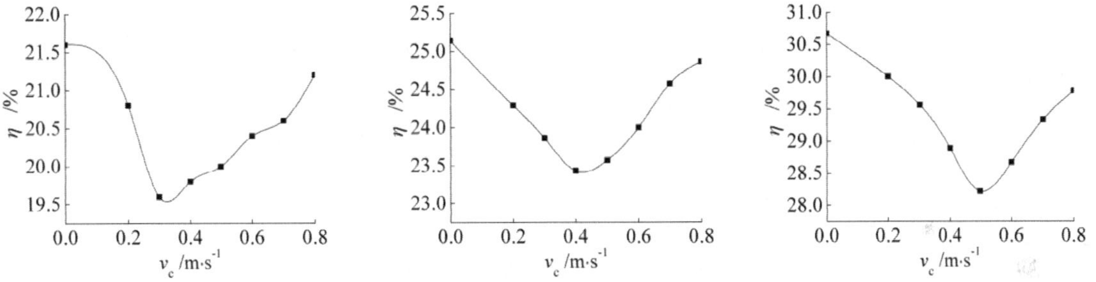

(a) t_1=40℃ **(b) t_1=50℃** **(c) t_1=60℃**

Figure 4. Crosswinds effect on cooling efficiency under different tower inlet water temperatures

The critical crosswind velocity v_{cr} corresponding to the operating conditions t_1=40 ℃, 50 ℃ and 60 ℃ is 0.3 m/s, 0.4 m/s and 0.5 m/s, respectively. It can be seen that the higher t_1 is, the bigger v_{cr} will become. This indicates that a higher t_1 makes it easier to eliminate the adverse effect of crosswinds. But on the other hand, the higher t_1 also can deteriorate the condenser performance. Therefore, in the actual operation of the cooling tower, it should first ensure the safe and economical operation of the condenser, on this basis, it can be appropriate to improve the tower inlet water temperature. If you want to eliminate crosswinds effect, you should seek other effective wind protection measures.

3.5. Inferences about the critical crosswind velocity and dimensionlessness of the crosswind velocity

Fr_c is defined as the crosswind Froude number, and that is $Fr_c = v_c / \sqrt{\Delta\rho g H_e / \rho}$. When a hot model test conforms to the kinematic similarity and thermo-dynamic similarity in the same time, the crosswind Froude number similarity will be satisfied naturally. According to the definition, Fr_c reflects the relative relationship between the crosswind inertia force and the tower draft force, which characterizes the crosswind effect on the cooling performance. Each Fr_c represents a cooling tower operating state under the effect of a crosswind, and the cooling tower operating state corresponding to v_{cr} is recorded as Fr_{cr}.

According to Fr_c similarity, v_c should proportionally vary with the tower draft force to keep the same cooling tower operating state. As mentioned above, when θ_1 decreases, Q and t_1 increases, the draft force of the cooling tower will increase, v_c will increase correspondingly, which is required to achieve the same air flow state in the tower. And that is the fundamental reason for the change of critical crosswind velocity.

The crosswind velocity v_c is dimensionless to the ratio of the crosswind velocity v_c to the air velocity v_f above the fill, denoted as the dimensionless crosswind velocity V_c. It can be found that η achieves minimum value when V_c=4 at all cases, and the cooling performance becomes worst, that is, the dimensionless critical crosswind velocity $V_{cr}= v_{cr}/v_f$=4. This conclusion has a strong theoretical and practical significance, that is, for a cooling tower, if you can get the air velocity above the fill, you can get the critical crosswind velocity corresponding to the worst cooling performance of the tower operating conditions.

4. Conclusion

In this paper, the crosswinds effect on the cooling performance of cooling tower is studied based on hot model test when the ambient temperature, circulating water flowrate and tower inlet water temperature changes. Specific conclusions are as follows:

1) The crosswind has a negative effect on the cooling tower thermal performance. There is a critical crosswind velocity v_{cr} corresponding to the lowest cooling efficiency, that is, when the crosswind velocity is lower than v_{cr}, the cooling performance will gradually reduce, and when the crosswind velocity is higher than v_{cr}, the cooling performance will gradually improve.

2) Cooling tower operation satisfies the crosswind Froude number similarity. When the ambient temperature decreases, the circulating water and the tower inlet water temperature increases, the cooling tower draft force will increase, the crosswind velocity will increase correspondingly, which is required to achieve the same flow state in the cooling tower. And that is the fundamental reason for the change of critical crosswind velocity.

3) At the dimensionless critical crosswind velocity V_c=4, the cooling efficiency η achieves the minimum value, and the cooling tower has the worst performance. For a cooling tower, if you can get the air velocity above the fill, you can get the critical crosswind velocity corresponding to the worst cooling performance of the tower operating conditions.

References

[1] Russell C M B, McChesney H R, Holder D W, et al. CROSS WIND AND INTERNAL FLOW CHARACTERISTICS OF DRY COOLING TOWERS[J]. 1978, 49(11): 20-24.

[2] Grange J L, Simon J Y. Behaviour of crossflow and natural draft cooling tower in the presence of wind[C]. IAHR Cooling Tower Workshop, Budapest, Hungary, 1982.

[3] Bender T J, Bergstrom D J, Rezkallah K S. Study on the effects of wind on the air intake flow rate of a cooling tower: Part 3. Numerical study[J]. Journal of Wind Engineering and Industrial Aerodynamics, 1996, 64(1): 73-88.

[4] Al-Waked R. Crosswinds effect on the performance of natural draft wet cooling towers[J]. International Journal of Thermal Sciences, 2010, 49(1): 218-224.

[5] Gao Ming, Sun Fengzhong, Wang Kai, et al. Experimental research of heat transfer performance on natural draft counter flow wet cooling tower under cross-wind conditions[J]. International Journal of Thermal Sciences, 2008, 47(7): 935-941.

[6] Zhao Yuanbin, Sun Fengzhong, Wang Kai, et al. Numerical analysis of crosswind effect on wet cooling tower aerodynamic field[J]. Nuclear Power Engineering, 2008, 29(6): 35-40.

[7] Zhou Lanxin, Jiang Bo, Chen Sumin. Numerical simulation for thermal performance of natural draft wet cooling tower[J]. Journal of Hydraulic Engineering, 2009, 40(2): 208-213.

[8] Chen Youliang, Sun Fengzhong, Wang Hongguo, et al. Experimental research of the cross walls effect on the thermal performance of wet cooling towers under crosswind conditions[J]. Applied Thermal Engineering, 2011, 31(17-18): 4007-4013.

IOP Conf. Series: Earth and Environmental Science **114** (2018) 012008 doi:10.1088/1755-1315/114/1/012008

Investigating the formation of acid mine drainage of Toledo pyrite concentrate using column cells

Diosa Marie Aguila

Batangas State University, Batangas City, Philippines 4200

dmmaguila@gmail.com

Abstract. Acid mine drainage (AMD) is an inevitable problem in mining and has adverse effects in water quality. Studying AMD formation will be valuable in controlling the composition of mine waters and in planning the rehabilitation method for a mine. In this research, kinetics of AMD formation of Toledo pyrite was studied using two column experiments. The mechanisms of AMD formation and the effects of various factors on pH drop were first studied. Another column test was done for validation and to study the role of Fe^{2+}/Fe^{3+} ratio in the change of leachate pH. The first experiment revealed that time and particle size are the most significant factors. It was also observed that the sudden pH drop during the starting hours was due to cracks formed from beneficiation, and the formation of $Fe(OH)_3$. The laddered behavior of pH thereafter was due to decrease in formation of $Fe(OH)_3$, and the precipitates in pyrite surface that lowered the surface area available for pyrite oxidation. The results of the second experiment validated the laddered behavior of pH. It was also observed that particle size distribution and pyrite surface were affected by the change in pH. Fe^{2+}/Fe^{3+} ratio of leachate generally decreased as pH dropped.

1. Introduction

The Philippines is known for its vast mineral resources. According to the Mines and Geosciences Bureau (MGB), the country has 9 million hectares of land identified with high mineral potential. One of the most important base metal productions of the country is copper, amounting to almost 16 billion pesos for January to September of 2013. Along with this, considerable amount of mine tailings are also produced and needed to be monitored. To prevent underestimation of the effects of tailings and other waste materials of mining in rehabilitation especially in the formation of pit lakes, extensive and continuous studies should be done.

One of the extensively studied effects of this is acid mine drainage (AMD), which is produced when, in a mine site, a sulfide mineral is exposed to oxygen and water, creating an acidic environment. It also releases high levels of metals that may have an impact to nearby land and water bodies and their biota [1].

Pyrite oxidation has been regarded as the most acid-producing sulfidic process in nature, thus being the primary source of AMD [2]. With its cyclic nature, this process can occur indefinitely.

Pyrite may be oxidized directly by oxygen or indirectly by oxygen and iron. Indirect oxidation starts with the directly oxidizing pyrite with oxygen, producing aqueous ferrous ion, sulfate and hydrogen ion. Then, ferrous ions released will be oxidized to ferric ions. This occurs at a low pH. The ferric ions along with oxygen will then oxidize pyrite, and constitute the process called the AMD engine, which will continue to occur until there's no pyrite left to oxidize. At pH greater than 3.0, Fe3+ reacts with water to form ferric (oxy)hydroxide. This reaction decreases Fe^{3+} in the solution, forming yellowish precipitates, $Fe(OH)_3$ [3].

However, the range of iron hydroxide/oxyhydroxide and sulfate minerals formed during pyrite oxidation is wider, depending on the amount of factors affecting the oxidation.
In view of the above, it can be reckoned that monitoring AMD and acid generating minerals exposed in the mine site as the mining progresses will not only help meeting stringent environmental requirements and regulations of the government but will also make mine rehabilitation more efficient.

According to Environmental Protection Agency (1994), there are two types of tests that are done to predict AMD: static and kinetic test. Static tests are usually done to know immediately the acid generating capacity of a mineral/rock at the mine site. Meanwhile, kinetic tests are done to predict the quality of water aiding the pyrite oxidation while manipulating the factors that may affect it. It is also done to have further information about the acid generating capacity of a mineral after doing static tests. Examples of kinetic tests are humidity cell test, soxhelet extraction test, and column tests.

Among these, column test is easily configured to suit the factors needed to be studied and has a relatively simple model. As such, many studies have been conducted using column test.

The objective of the study is to determine the kinetics of acid mine drainage formation using the pyrite samples obtained from Toledo Mines, Philippines.

2. Methodology

2.1. Analytical Methods

The pH of the leachate were measured using a pH meter (PH-501, General). The mass of the samples was measured using an analytical balance (ATX224, Shimadzu). Screen analysis was done using Rotap and Tyler sieves. The surface morphology of pyrite was analysed by scanning electron microscope (SEM) (XL 30, Philips).

2.2. Sample Preparation

Jones Riffle was used to get 1 kg representative sample. Pyrite was then sieved using ASTM mesh sizes #25, #70, #100, #140, #200, #270, #400 and #800 to get the particle size distribution.

Three particle ranges are used in the experiment: ≥149 μm (P1), ≤149 -105 μm (P2), and ≤105 μm (P3).

Samples were stored in airtight plastic bags until use to avoid surface oxidation.

2.3. Column Cells Setup

Twelve (12) Plexiglass columns (24 inch length, 1 inch inside diameter) were fabricated and were equipped with air pumps. For the design of the experiment, 3 particle ranges (P1, P2 and P3), 2 different water levels (50 mL and 100 mL) and 2 pyrite sample masses (5 g and 10 g) were use, as shown in Table 1.

The air injected in the setup is continuous and fixed throughout the experiment.

Table 1. Packing of samples in the column cell setups

Column No.	Particle Size	Water Level (ml)	Sample Mass (g)	Column No.	Particle Size	Water Level (ml)	Sample Mass (g)
1	P1	50	10	7	P2	50	5
2	P1	100	10	8	P2	100	5
3	P1	50	5	9	P3	50	10
4	P1	100	5	10	P3	100	10
5	P2	50	10	11	P3	50	5
6	P2	100	10	12	P3	100	5

Desired water quantity was poured to all columns and was let stand for an hour to let the water seep through the solids. Air was then introduced using air pumps (E-JET268 2.7L/min, 150 Pa). After an hour, first pH measurement was conducted. Second pH measurement was done after 6 hours. The succeeding measurements were done once a day for 30 days.

3. Results

3.1. Column Study: Mechanisms

Shown in figure 1 are the pH measurements of the leachates per column cell within 30 days. The sudden pH drop during the first 6 hours is due to the available surface cracks and irregularities brought about by the comminution process. These cracks served as reactivity points where it's easiest to oxidize pyrite [4]. The graphs also show that most of the columns with the lowest particle size (P3), thus having highest surface area available, have relatively lower pH than the other measurements.

Shown in figure 2 is a SEM image of the unreacted pyrite surface showing visible cracks and other surface irregularities.

The proposed governing reactions in that phase of fast pH drop are the following:

$$FeS_2 + 7/2\ O_2 + H_2O \rightarrow Fe^{2+}_{(aq)} + 2\ SO_4^{2-}{}_{(aq)} + H^+_{(aq)} \quad (1)$$

$$4\ Fe^{2+}_{(aq)} + O_2 + 4\ H^+_{(aq)} \rightarrow 4\ Fe^{3+}_{(aq)} + 2\ H_2O \quad (2)$$

$$Fe^{3+}_{(aq)} + 3\ H_2O \leftrightarrow Fe(OH)_3 + 3\ H^+ \quad (3)$$

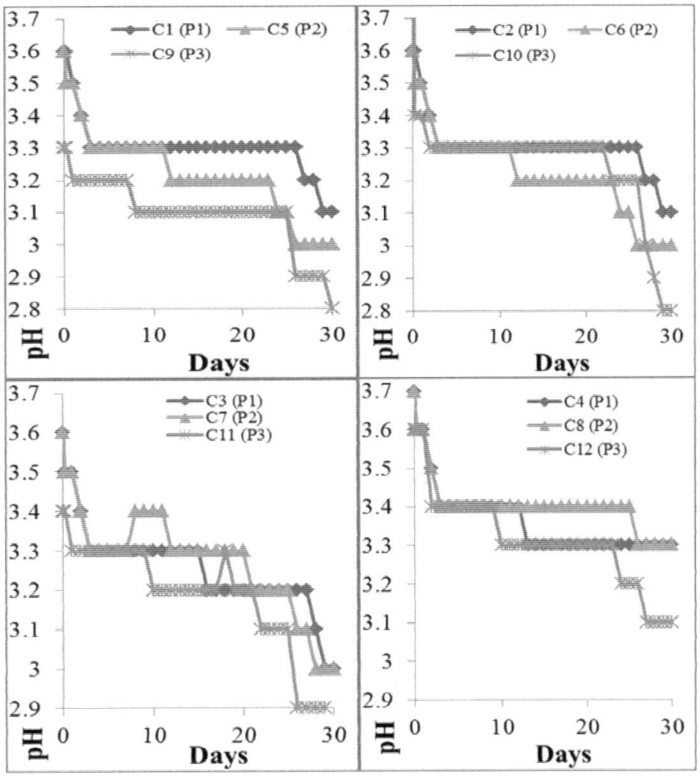

Figure 1. pH measurements within 0-6 hours for leachate samples in C1, C5 and C9 (top left), C2, C6 and C10 (top right), C3, C7 and C11 (bottom left) and C4, C8 nd C12 (bottom right)

Pyrite in contact with water and oxygen produced ferrous ion (Eq. 1). Since the system had adequate oxygen supply by means of the air pump, ferrous ions were oxidized to ferric ions (Eq. 2). At pH above 3.5, ferric ions are precipitated into amorphous ferric hydroxide (Eq. 3) which released more hydrogen ions. Thus, the fast drop of pH during the first contact of pyrite, water and oxygen was due to cracks which increased the surface available for oxidation and formation of $Fe(OH)_3$.

On the other hand, after one or two days, the leachates' pH in all column cells decreased slowly, in a laddered manner, as shown in Fig. 2. In this stage, the columns with the smallest particle size also have lower pH at the end of the experiment.

As pH decreased from 5.6 – 3.5, some of the ferric hydroxide formed dissolved back to ferric ions because of higher solubility of ferric hydroxide at lower.

This reaction consumed hydrogen ions thus effectively buffering the pH of the leachate. Also, the oxidation of ferrous ions to ferric ions became repressed at pH below approximately 4.5, thus $Fe(OH)_3$ and H^+ production became repressed, too [4,5]. This means that the reaction governing this phase of slow pH decrease is the oxidation of pyrite into ferrous ions.

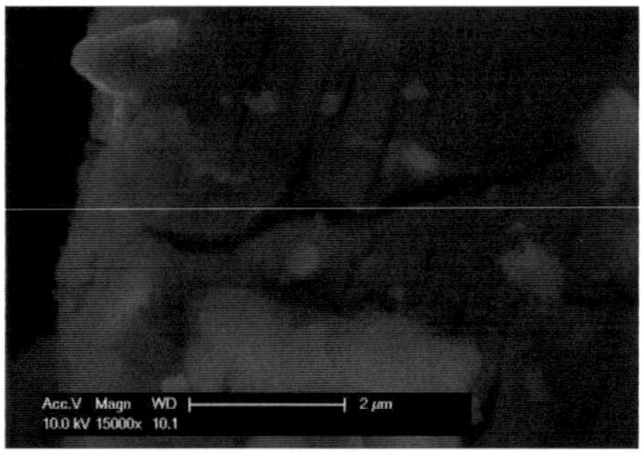

Figure 2. SEM image of unreacted pyrite surface showing cracks

3.2. Column study: Effects of factors

The effects of factors such as time, particle size, mass of samples, and volume of leachate. The levels per variable are shown in Table 2. Design Expert was used for the analysis of the results.

Mean interaction plots were used to know the effect of the factors in the variation of the result and their interactions in the experiment. The plots in figure 3 show that the most significant factor and causes higher variation in the results, the one with the steepest slope, is the time. Next most significant factor is particle size . In the previous literatures that studied effects of particle size in pH, finer particle size results to lower pH [6,7], which is consistent with the results in this study.

Table 2. Level of factors

Parameter	-1	0	1
Time (day)	1	15	30
Mass (g)	5	-	10
Volume (ml)	50	-	100
Particle Size	P3	P2	P1

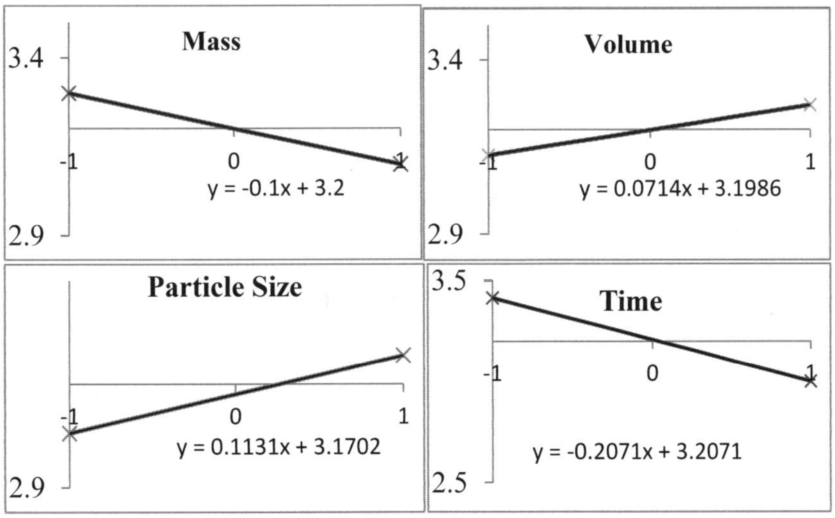

Figure 3. Mean interaction plots for pH of the leachate based on the factors

4. Conclusion

1. The formation of acid mine drainage in Philippine pyrite in column cell experiment was initiated by the formation of $Fe(OH)_3$ precipitates which increased the formation of hydrogen ions. The fast decrease of pH at the start was due to the cracks formed from the beneficiation process that increased the surface area of the pyrite available for reaction. A gradual decrease in pH afterwards until the end of experiment was governed by the decrease in rate of oxidation of pyrite due to the precipitates in the pyrite surface that lowered the surface area available for pyrite oxidation.
2. Among the factors, contact time and particle size showed the most influence in pH.

Acknowledgments

The author would like to thank Engineering Research and Development for Technology (ERDT) for supporting the financial aspects of this endeavour.

References

[1] Lowson R 1982 Aqueous Oxidation of Pyrite by Molecular Oxygen *Chemical Reviews 82*, pp 461–497
[2] Chandra A and Gerson A 2010 The mechanisms of pyrite oxidation and leaching: A fundamental perspective *Surface Science Reports 65*, pp 293-315
[3] Evangelou V 1995 *Pyrite oxidation and its control* (Florida: CRC Press, Inc.)
[4] Lottermoser B 2007 *Mine Wastes: Characterization, Treatment and Environmental Impacts* (Heidelberg: Springer)
[5] Barik R and Mohapatra M 2011 Facile precipitation of iron oxide with in situ Fe(II) in $Fe_2(SO_4)^{3-}NaOH^-(N_2H_4)_2H_2SO^{4-}H_2O$ medium *Indian Journal of Chemical Technology* vol 18 pp 107-112
[6] Butler B 2009 Effect of pH, ionic strength, dissolved organic carbon, time, and particle size on metals release from mine drainage impacted streambed sediments *Water Research 43* 1392 – 1402
[7] Erguler G Erguler Z Akcakoca H and Ucar A 2014 The effect of column dimensions and particle size on the results of kinetic column test used for acid mine drainage (AMD) prediction *Minerals Engineering 55* pp 18-29

Research on Precision Tracking on Fast Steering Mirror and Control Strategy

Lin Di[1,2], Wu Yi-ming[1] and Zhu Fan[1]

[1] Xi'an Institute of Optics and Precision Mechanics, Chinese Academy of Sciences, 710119,China
[2] University of Chinese Academy of Sciences,Beijing 100049,China
E-mail: lindi2017@tom.com

Abstract. Fast steering mirror is a device used for controlling the beam direction precisely. Due to the short travel of the push-pull FSM, a compound fast steering mirror system driven by both limited-angle voice coil motor and push-pull FSM together is proposed. In the compound FSM system, limited-angle voice coil motor quickly swings at wide angle, while the push-pull FSM do high frequency movement in a small range, which provides the system with the high bandwidth and long travel. In the control strategy, the method of combining feed-forward control in Kalman filtering with auto-disturbance rejection control is used to improve trajectory tracking accuracy. The simulation result shows that tracking accuracy measured by the compound method can be improved by more than 5 times than that of the conventional PID.

1. Introduction

Precision capture, tracking and targeting systems play a central role in large-size photoelectric measurement equipment, astronomical observation and laser communication. For example, high-accuracy stabilized sighting pod and IRSI have high requirement on stabilization precision of sighting line. Most of the current photoelectric stabilized sighting systems adopt the overall stability mode. Due to the effects of load and various disturbance torque, the overall stabilization precision is usually only 25 ~ 30urad. Therefore, the use of fast steering mirror and compound-axis structure is the effective means for stabilization with large load and high accuracy [1] ~ [6].

At present, drive modes of FSM based on voice coil motor mainly include push-pull, limited-angle, etc., among which push-pull voice coil motor has advantages of fast response rate, high bandwidth, low power consumption, high displacement accuracy, etc., but its travel is short. The bandwidth of limited-angle voice coil motor is lower than that of the piezoelectric ceramic, while its travel is relatively long. The existing FSM is basically driven by independent push-pull voice coil motor or limited-angle voice coil motor, but it does not have the advantages of long travel and high bandwidth at the same time[7].

This paper presents a compound fast steering mirror structure, as shown in Fig. 1, in which first-level steering mirror presents X-Y structure, and is driven by limited-angle voice coil motor, resulting in large range of angle and small shaft coupling. The secondary push-pull FSM has the advantages of high resolving power, fast response rate, etc. Compound fast steering mirror has the advantages of large travel and high bandwidth at the same time. Based on this structure, simulating evaluation and tracking performance on the method of combining ADRC of FSM with compound control with feed-forward compensation are conducted.

2. Principle of Servo Control

IOP Conf. Series: Earth and Environmental Science **114** (2018) 012009 doi:10.1088/1755-1315/114/1/012009

The compound fast steering mirror is composed of quick-swing outer frame driven by voice coil motor and internal push-pull FSM unit. The outer frame has a large range of swing angle, which can realize quick angular positioning. The internal fast steering mirror unit based on piezoelectric ceramic can realize fast image stabilization and beam stabilization in a small angular range. Configuration of limited-angle voice coil motor is as shown in Fig. 2.

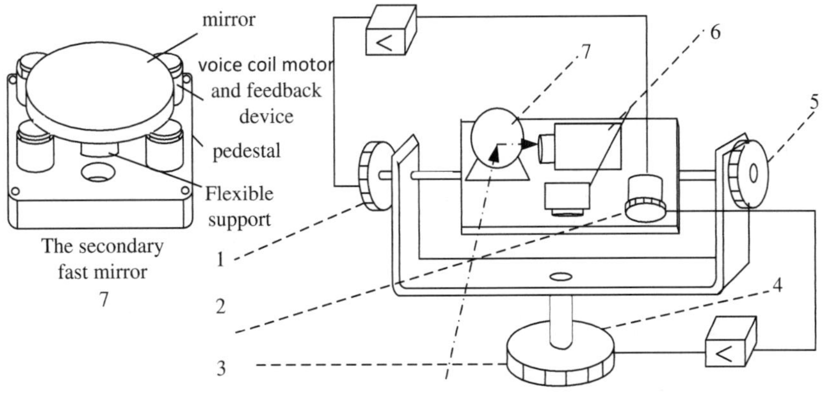

1、Pitch voice coil motor 2、Gyroscope 3、Azimuth sensor

4、Azimuth voice coil motor 5、Pitch sensor 6、Camera

7、Secondary fast mirror

Figure 1. Structure schematic diagram of compound fast steering mirror.

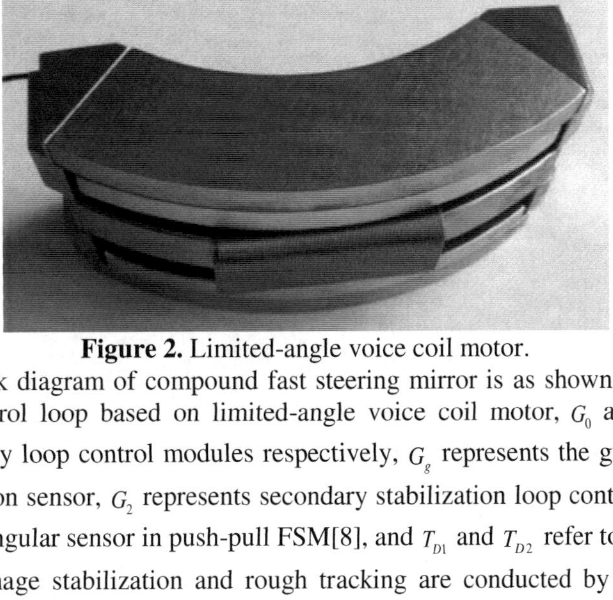

Figure 2. Limited-angle voice coil motor.

The functional block diagram of compound fast steering mirror is as shown in Fig. 3, in which the main loop is a control loop based on limited-angle voice coil motor, G_0 and G_1 represent the position loop and velocity loop control modules respectively, G_g represents the gyroscope sensor, G_θ represents angular position sensor, G_2 represents secondary stabilization loop control module in push-pull FSM, G_p refers to angular sensor in push-pull FSM[8], and T_{D1} and T_{D2} refer to disturbing terms.

In Fig. 3, rough image stabilization and rough tracking are conducted by first-level platform driven by the limited-angle voice coil motor, secondary image stabilization and precision tracking are performed by fast steering mirror driven by the push-pull voice coil motor.

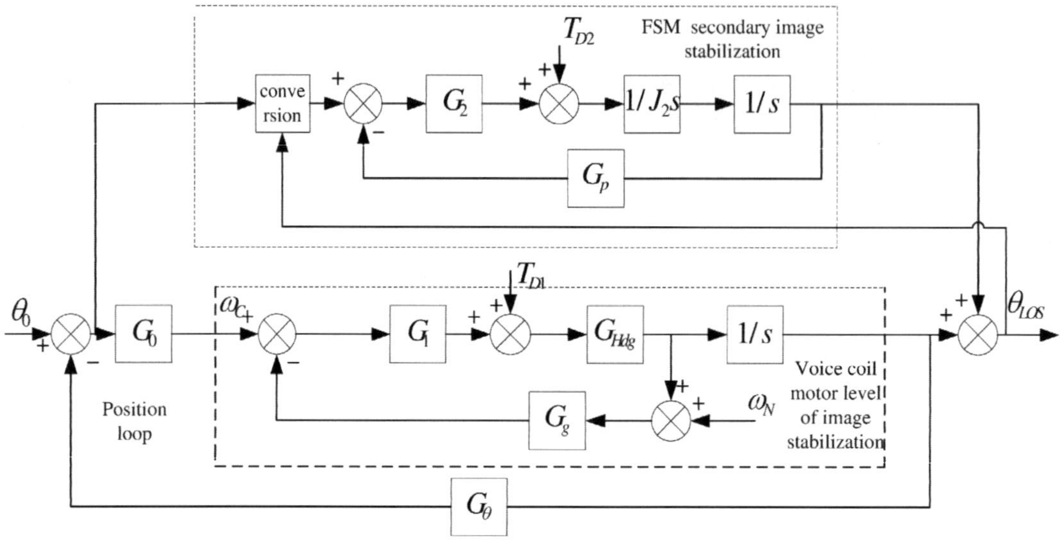

Figure 3. The functional block diagram of compound fast steering mirror.

3. Strategy on Combining ADRC with Feed-forward Control of Closed-Loop System in Compound Fast Steering Mirror

As shown in Fig. 4, the push-pull FSM module in compound fast steering mirror adopts the mode of combining ADRC with feed-forward control. Kalman filtering is used in feed-forward control to filter angular, and to estimate angular velocity and angular acceleration signals so as to form feed-forward compensation control.

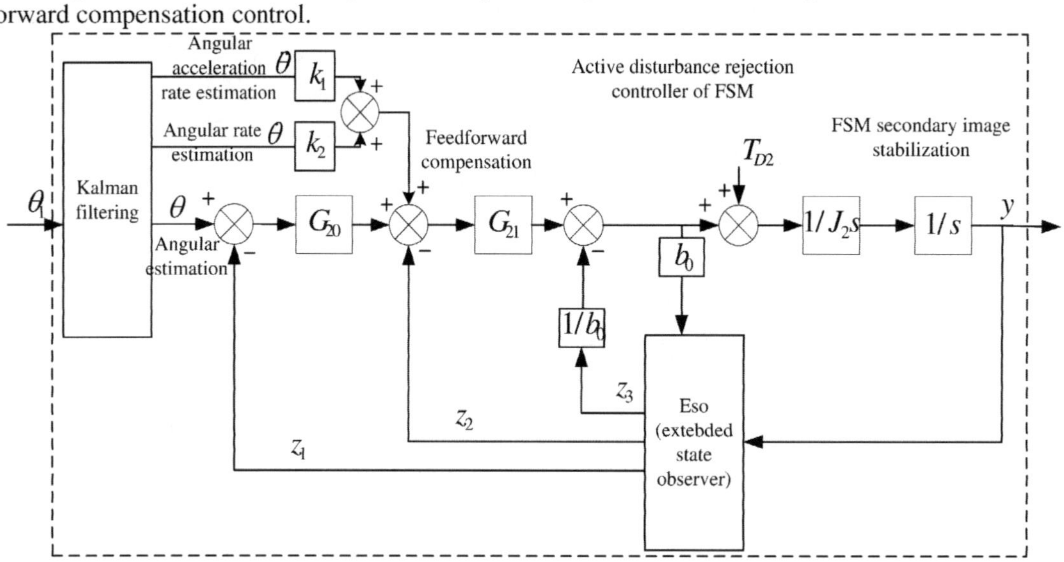

Figure 4. Design on combining ADRC of FSM in tracking mode with feed-forward control.

ADRC is used for observing the external disturbance and providing feed-back compensation, and feed-forward compensation is used for enhancing tracking accuracy of the high-speed galvanometer on fast signal. In ADRC based on piezoelectric ceramic model, as shown in Fig. 4, nonlinear extension state observer designed by Han Jingqing shows [9]:

$$\begin{cases} e = z_1 - y \\ \dot{z}_1 = z_2 - \beta_1 e \\ \dot{z}_2 = z_3 - \beta_2 fal(e, \alpha_1, \delta) + bu \\ \dot{z}_3 = -\beta_3 fal(e, \alpha_2, \delta) \end{cases} \tag{1.1}$$

Where, $\beta_i > 0(i=1,2,3)$, $\alpha_1 = 1$, $\alpha_2 = 0.5$ 。

The function of saturation function $fal(e,\alpha,\delta)$ is to suppress the chattering of signal. It can be expressed as follows:

$$fal(e,\alpha,\delta)=\begin{cases}\dfrac{e}{\delta^{1-\alpha}} & |e|\le\delta \\ |e|^{\alpha}\operatorname{sgn}(e) & |e|>\delta\end{cases} \tag{1. 2}$$

Simplifying the transfer function of FSM and converting it into a differential equation, we can get the solution:

$$\begin{cases}\dot{x}_1 = x_2 \\ \dot{x}_2 = -6849x_2 + 6849u + f(x_1,x_2) \\ y = x_1\end{cases} \tag{1. 3}$$

Where, $f(x_1,x_2)$ remains unknown.

When $f(x_1,x_2)=33\sin(1000\pi t)$ and $b=6849$, extension state observer is used for observing the unknown part to conduct simulation of control system.

Simulation efficiency and errors of trajectory tracking by the conventional PID control are as shown in Fig. 5. Trajectory tracking and error simulation efficiency from combining ADRC with feed-forward control are as shown in Fig. 6.

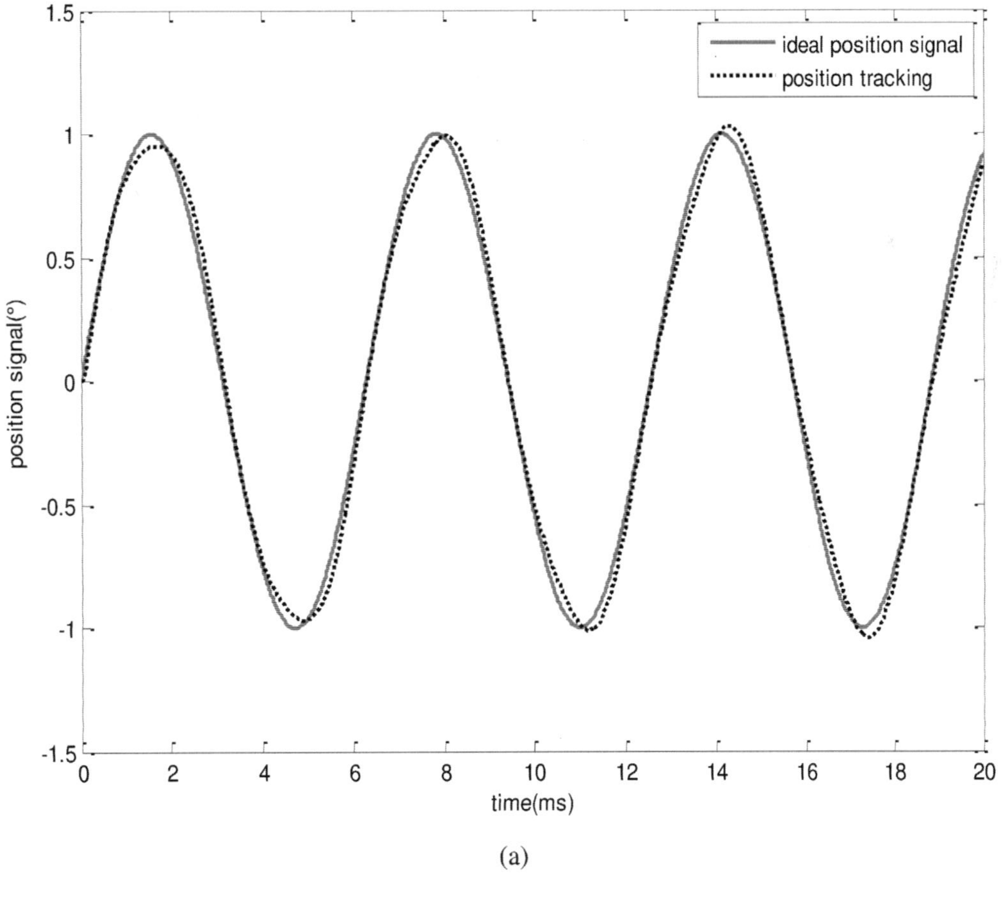

(a)

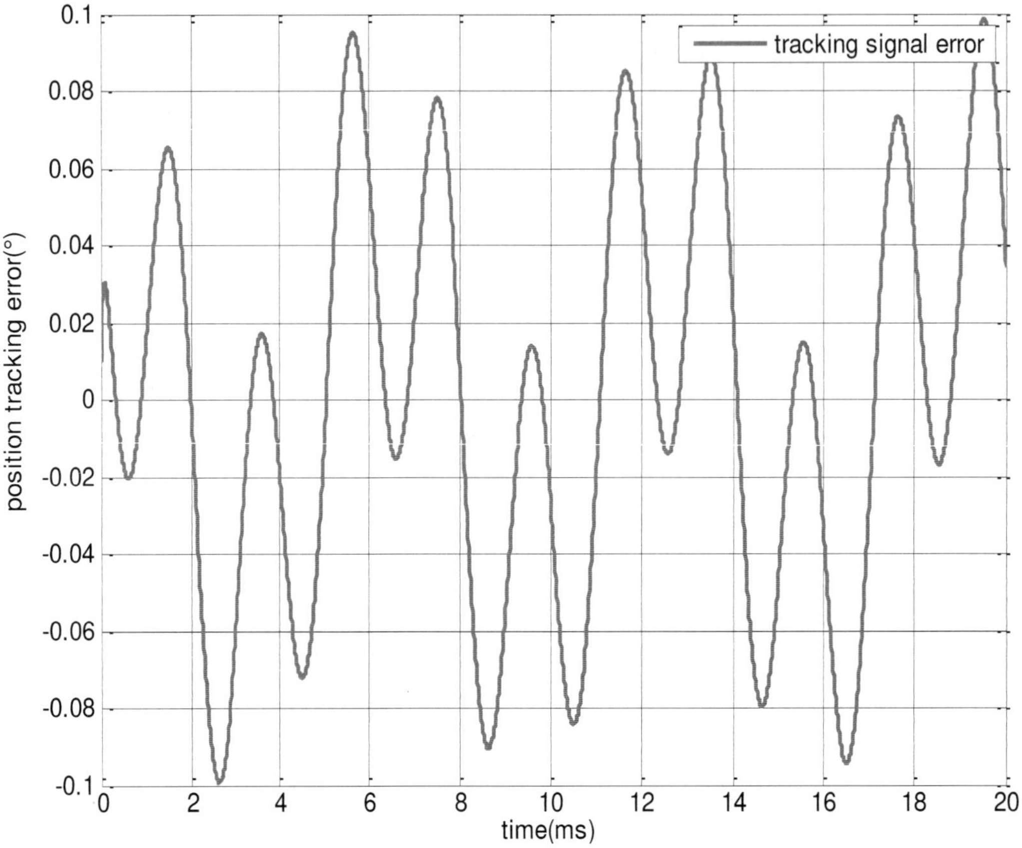

(b)

Figure 5. Continuous trajectory tracking and tracking error from conventional PID control.

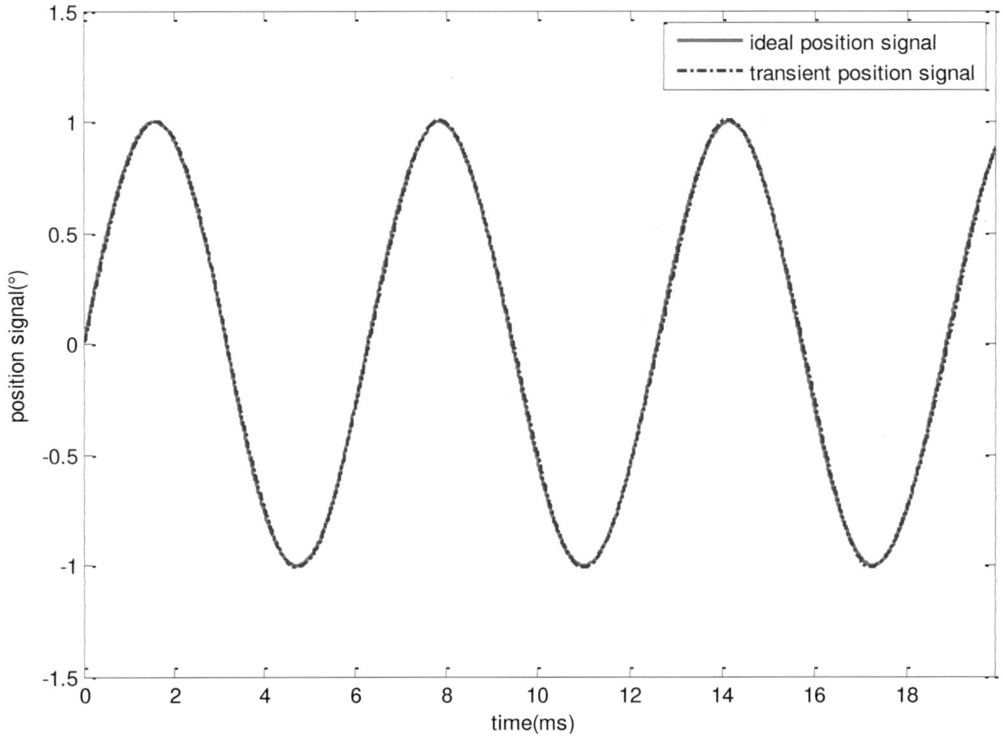

(a)

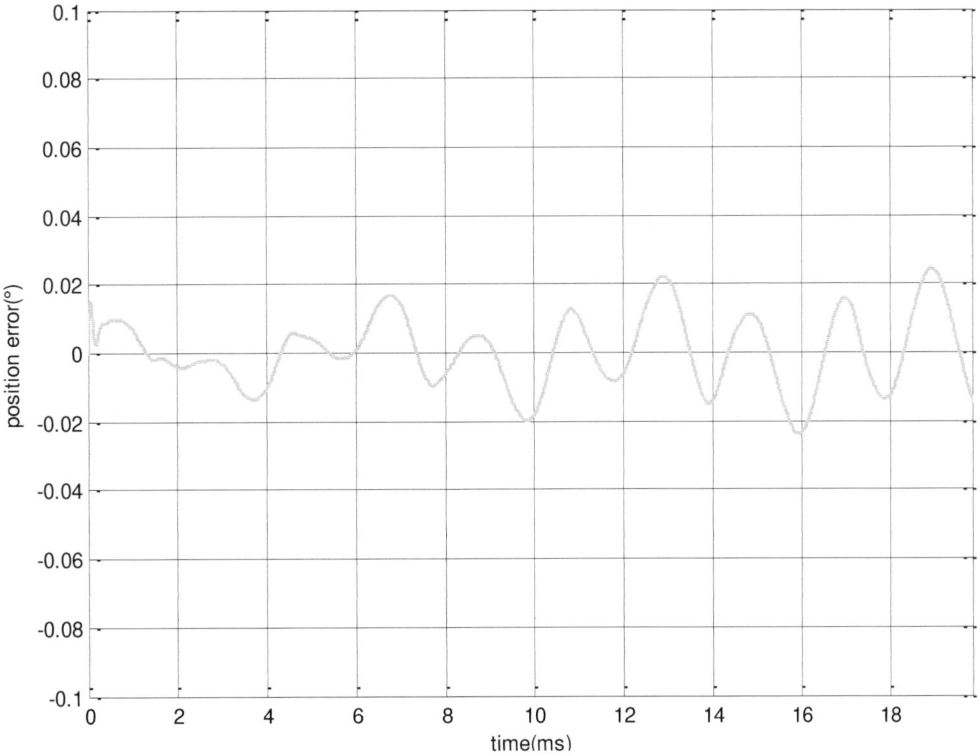

(b)

Figure 6. Trajectory tracking and tracking error of combining ADRC with feed-forward control.

Compared Fig. 5 with Fig. 6, we can see that the conventional PID control produces relatively large tracking error due to the presence of external disturbance signal and control delay. Combining ADRC with feed-forward control can estimate the external disturbance and provide compensation, on the other hand, its estimated velocity and acceleration can be used as feed-forward control to improve trajectory tracking accuracy. From the simulation results, we can see that in the case of the same external disturbance, the algorithm in this paper reduces the comprehensive control error of the system by a factor of 5, with strong capacity of resisting disturbance and good control performance. Integrated position, velocity and disturbance estimation of ADRC are as shown in Fig. 7, in which the expansion state observer can accurately estimate the feedback velocity signal and the disturbance signal, and correct velocity loop and provide disturbance compensation in the control loop.

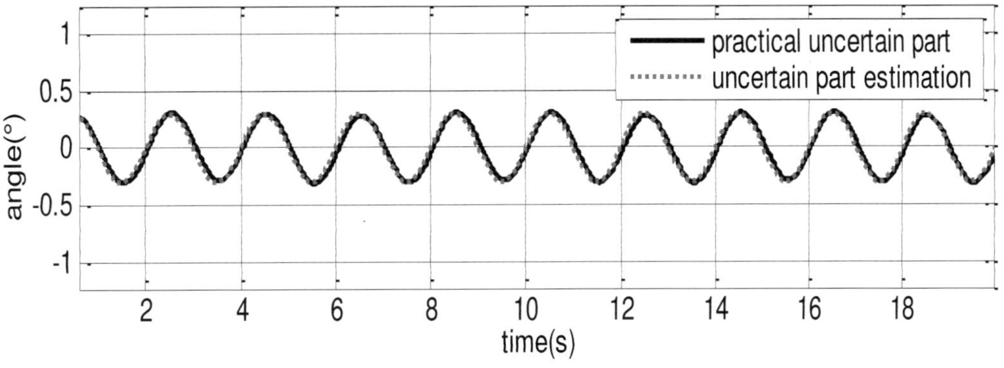

(a)

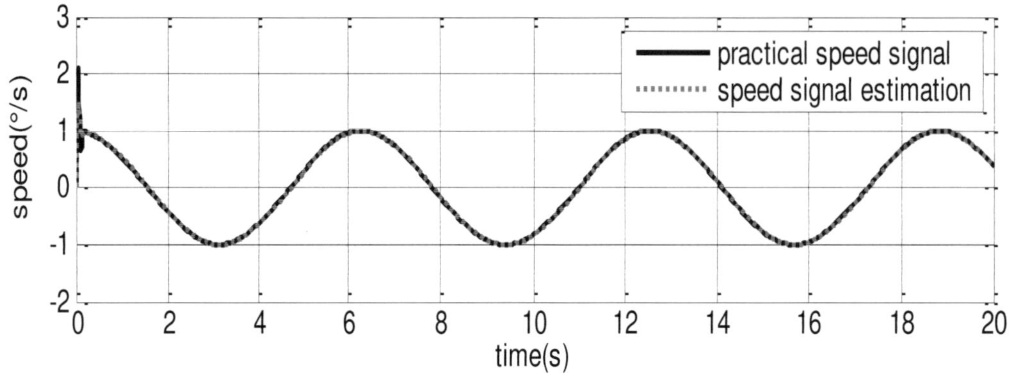

(b)

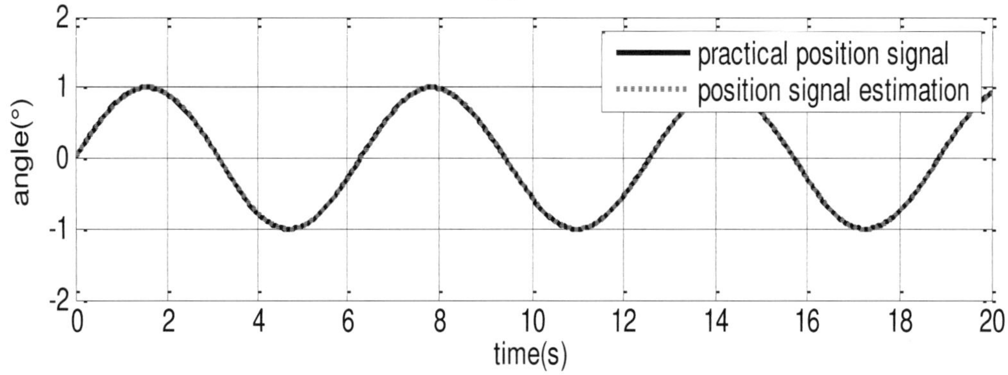

(c)

Figure 7. Observed results from expansion state observer.

4. Conclusion

In order to compensate for the shortcoming of short travel of piezoelectric ceramic galvanometer, limited-angle voice coil motor is introduced to drive the galvanometer frame of piezoelectric ceramic to swing quickly at large angle, which constitutes compound fast steering mirror with a large travel, high bandwidth and high accuracy. In the control strategy, the method of combining ADRC with feed-forward compound controller is introduced so that velocity loop control, feed-forward control and disturbance compensation control are added on the basis of positioning control in the conventional piezoelectric ceramic. From the simulation results, we can see that its trajectory tracking accuracy get greatly improved than that of the conventional PID control.

5. References and Citations

[1] Peng Shuping and Yu Hongjun. Design of servo system for novel fast-steering mirror [J]. *Infrared and Laser Engineering*, 2014, 43(5):1610-1615. (in Chinese)

[2] Wang Heqi and Wang Weiguo. Solution of correcting values on laser beam directing of fast steering mirror for mobile vehicle [J]. *Optics and Precision Engineering*, 2016, 24(6):1399-1406. (in Chinese)

[3] CHOI Y J and KANG M S. Dual stage servo controller for image tracking system [J]. *Journal of the Korean Society of Precision Engineering*, 2007, 24(2):86-94.

[4] HILKER J M. Inertially stabilized platform technology [J]. *IEEE Control systems Magazine*, 2008, 2:26-46.

[5] Wen Jianghua and Zhao Chuangshe. Study of Driving and Controlling with Fast Steering Mirror Based on PZT [J]. *Ordnance Industry Automation*, 2015, 34(8):59-64. (in Chinese)

[6] Sun Chongshang. Research on the Scanning Image Motion Compensation Technology Based on Fast Steering Mirrors with High Precision and Wide Frequency Range [D]. BeiJing: University of Chinese Academy of Sciences, 2016. (in Chinese)
[7] Xu Xinhang and Yang Hongbo. Research on key technology of fast-steering mirror [J]. *Laser and infrared*, 2013, 43(10):1095-1103. (in Chinese)
[8] Xu Feifei and Liu Sha. Performance comparison and analysis of coarse and fine combined stabilization control system based on mirror compensation [J]. *Applied optics*, 2013, 34(1):15-20. (in Chinese)
[9] Qiao Guanyu. Research on Active Disturbance Rejection Control Strategy of Fast-Steering Mirror Closed Loop System [D]. Beijing: University of Chinese Academy of Sciences, 2012. (in Chinese)

IOP Conf. Series: Earth and Environmental Science **114** (2018) 012010 doi:10.1088/1755-1315/114/1/012010

A Multipath Mitigation Algorithm for vehicle with Smart Antenna

Jing Ji[1,2], Jiantong Zhang[2], Wei Chen[3] and Deliang Su[1]

[1] School of Information Engineering, Wuhan University of Technology, Wuhan, CHINA
[2] Satellite Navigation Division of CTTIC, Beijing, CHINA
[3] School of Automation, Wuhan University of Technology, Wuhan, CHINA
E-mail: 362401773@qq.com

Abstract. In this paper, the antenna array adaptive method is used to eliminate the multipath interference in the environment of GPS L1 frequency. Combined with the power inversion (PI) algorithm and the minimum variance no distortion response (MVDR) algorithm, the anti-Simulation and verification of the antenna array, and the program into the FPGA, the actual test on the CBD road, the theoretical analysis of the LCMV criteria and PI and MVDR algorithm principles and characteristics of MVDR algorithm to verify anti-multipath interference performance is better than PI algorithm , The satellite navigation in the field of vehicle engineering practice has some guidance and reference.

1. Introduction

GNSS, global satellite navigation system. Four active systems are existed as: US-led GPS, Russia-led GLONASS, China-led BDS and European-led GALILEO. With the development of satellite navigation applications, GPS has become the world's most trusted users, the most important spatial information infrastructure, GPS system to provide users with the whole period, all-weather positioning, navigation and timing (PNT) service capabilities have become the most important national capacity in the United States. In recent years, the deteriorating electromagnetic environment and man-made threats have caused signal transparency and interface to open the GNSS system to further increase the vulnerability, making the receiver facing difficult to meet the military needs, homeland security, economic development, infrastructure and people's daily life on the positioning, Navigation and timing requirements of the dilemma, the urgent need to strengthen the receiver anti-jamming and interference cancellation research.

The most common interference scenarios for GNSS receivers in civil applications are multipath disturbances, which occur in urban canyons and tall mountains, and the GNSS receivers receive signals that are formed by mixing the direct and reflected portions, In extreme cases, only the reflected part of the received signal, multi-path interference caused by the GPS positioning accuracy of the maximum error even up to 150 meters. According to the principle of division, in the receiver section of anti-jamming technology has three: antenna technology, receiver technology, post-processing technology [1]. The basic principle of the antenna technology is to compare the desired line of sight signal direction vector and the actual signal direction vector difference to carry out multi-path detection [2]. It has the advantage that the adjustment factor can be used to suppress multi-path interference accuracy, but the cost is higher; receiver technology is through pre-detection bandwidth,

narrowed signal code tracking loop filter and carrier tracking loop filter bandwidth approach to improve Anti-jamming performance, the advantage is that the external navigation system can provide accurate speed to make up for the loss of the dynamic range, the disadvantage is that the code / carrier tracking ring can not be long lost [3]. The post-processing technology is also called fusion technology, mainly through the Kalman filter algorithm and other related algorithms to integrate multi-sensor, multi-system solution data to correct the positioning error caused by multipath interference, which has the advantage of achieving high precision positioning of the receiver , The disadvantage is that real-time elimination of interference can not be achieved [4].
The antenna array is composed of multiple antennas, and its essence is to use the spatial array of sensor arrays and multi-channel reception to obtain the multi-dimensional information such as time domain and airspace of the signal source to achieve the purpose of detecting the signal and extracting its parameters. in fact. Present form of adaptive antenna technology, antenna array beam formation of the satellite signal to enhance the receiving technology, zero antenna technology, vector antenna technology. These methods can not only inhibit the influence of multipath signals. And can suppress narrowband or broadband interference. To achieve the signal to the wave direction identification and signal polarization diversity reception.
The rest of this paper is organized as follows. Section 2 introduces the signal model and Section 3 the mitigation algorithm is compared for the vehicle. In section 4, some results are presented to evaluate in the vehicle test, and conclusions are drawn in Section 5.

2. Signal Model

2.1. Multipath Inference
Improvements due to GNSS augmentations and GNSS modernization are reducing many sources of error, leaving multipath and shadowing as significant and sometimes dominant contributors to error, as shown in Figure 1.This section discusses these sources of error, their effects, and ways to mitigate their effects [5].

Figure 1. Multipath Effect

Let's assume we receive total combined signal x(t), it can be decomposed into three distinct components: desired signal, interference signal, and the noise. Thus, x(t) can also is described in three components form as

$$x(t) = x_d(t) + x_i(t) + x_n(t) \tag{1}$$

Where the subscripts *d*, *i* denote the desired signal, interference, and $x_n(t)$ is the thermal noise of propagation.
For the desired signal $x_d(t)$ with no multipath and noise is given by

$$x_d(t) = A_0 C(t - \tau_0) e^{-j\phi_0} e^{j2\pi f_c \tau_0} \tag{2}$$

Where A_0 is the received amplitude of the desired signal, $C(t)$ is the complex envelope of the transmitted signal, τ_0 and ϕ_0 are the propagation delay and carrier phase of the desired signal from satellite to receiver, and f_c is the carrier frequency. Then the multipath interference with the same form is expressed as

$$x_i(t) = \sum_{i=1}^{L} A_i C(t - \tau_i) e^{-j\phi_i} e^{j2\pi f_i \tau_i} \tag{3}$$

Where there are L multipath, A_i is the receiver amplitude of the ith multipath interference, τ_i is the propation delays of the ith multipath returns, ϕ_i is the carrier phases of ith multipath returns. and f_i is the received frequencies of the ith multipath return relative to the carrier frequency.
The expression (1) can be rewritten using parameters with neglecting noise $x_n(t)$ after sampling as

$$x(t) = A_0 e^{-j\phi_0}\left[C(t - \tau_0) + \sum_{i=1}^{L} \widetilde{A_i} e^{-j\widetilde{\phi_i}} C(t - \tau_0 - \widetilde{\tau_i})\right] \tag{4}$$

Where $\widetilde{A_i}$ denote the ith interference to desired ratio of amplitude, $\widetilde{\tau_i}$ is the relative delay of the ith multipath returns, and $\widetilde{\phi_i} = \phi_i - \phi_0$ is the relative carrier phases with ith multipath returns.
The expression (4) implies that the receive signal consist of a relatively stable direct signal(neglect shadowing occurs) and multipath interference drastically with changes in circumstance reflections. The change in the received signal is more dependent on the relative variable. Theoretical analysis, the desirable signal can be captured and tracked by technique with relative amplitude weights, relative delays and relative phases, such as DLL(Delay Lock Loop) and PLL(Phase Lock Loop), and similarly, the interference can be eliminated with waveform distortionless algorithm in delay and phase [6] .

2.2. Adaptive Antenna Array

Adaptive antenna array, also known as Smart antenna, is composed of the antenna array, ADC (Analog to Digital Converter) and baseband processing part. According to the array structure arrangement, it can be divided into: ULA (Uniform Linear Array), UPA (Uniform Planar Array), UCA (Uniform Circular Array) and spherical array.
The intelligent antenna can adaptively adjust the maximum radiation direction of the beam, and track the desired signal dynamically, and align the main signal with the desired signal, the null point to the interference signal, and the maximum suppression of the interference, using the signal antenna and the phase difference signal[7].

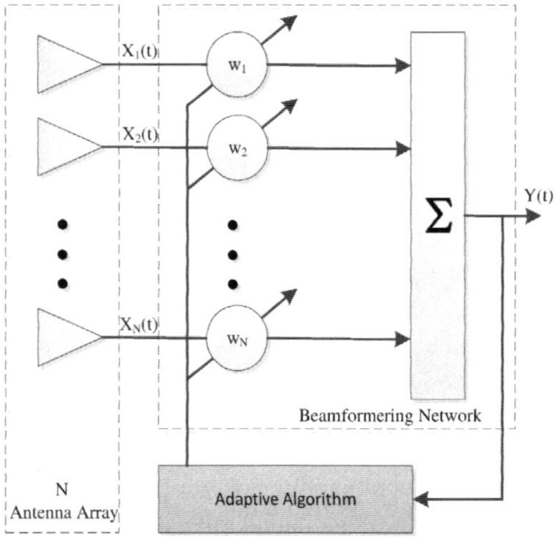

Figure 2. Adaptive Antenna Arrays Model

For a uniform array with N elements, d is denoted the spacing between adjacent elements, the signal received by each array through the adaptive processing module, get the N-way weight, the signal data and the corresponding right Value is weighted to get the combined output [8], as shown in Figure 2. The angle of incidence of the expected signal is θ and the elevation angle is φ. It can be expressed as

$$y_n(\theta,\phi) = \vec{W}^H \cdot \vec{X}(n) \tag{5}$$

The received signals array

$$X(t) = AS(t) + N(t) \tag{6}$$

$X(t) = [X_1(t),\cdots,X_N(t)]^T$ is data vector with N elements array, $N(t) = [N_1(t),\cdots,N_N(t)]^T$ denote noise vector, $S(t) = [S_1(t),\cdots,S_N(t)]^T$ is the received signal complex envelope vector, $S_p(t)$ is the complex of the pth signal source Envelope. $A = [a(\theta_1),\cdots,a(\theta_m)]$ denote steering matrix, $a(\theta_p)$ denote the steering vector with pth source, and $a(\theta_p) = [1, e^{j\beta_p},\cdots,e^{(N-1)j\beta_p}]$, (P=1,2, ⋯m)

$$\beta_p = \frac{2\pi}{\lambda} d\sin\theta_p \tag{7}$$

Define the array covariance matrix

$$R = E[X(t)X^H(t)] \tag{8}$$

The weight is denoted by

$$W = [W_1,\cdots,W_N] \tag{9}$$

Then the array signal is weighted after the synthesized power output as

$$P_{out} = E\{|y(n)|^2\} = E\left\{\left(W^H X(n)\right)\left(W^H X(n)\right)^*\right\} E\{W^H R_{xx} W\} \tag{10}$$

3. Beamforming algorithm

3.1. PI

PI(Power Inversion) algorithm based on the LCMV criterion, so that the response vector g is constant 1, Constraint matrix $C = S = [1,0,\cdots,0]^T$, the linear constraint condition as follow

$$W^H S = 1 \tag{11}$$

That is, the constraint factor for the first branch of the weighting factor is always 1, so that the output power is minimum, PI algorithm optimal weight vector is calculated as

$$W_{PI} = R_{XX}^{-1} S\left(S^H R_{XX}^{-1} S\right)^{-1} \tag{12}$$

The signal after interference suppression is

$$S_{PI}(n) = W_{PI}^H\, a(\theta_p)s(n) \tag{13}$$

3.2. MVDR

MVDR (Minimum Variance Distortionless Response) algorithm was first proposed by Capon in 1967, also known as Capon algorithm, which is an adaptive spatial beam spectrum estimation algorithm. Based on the linear minimum criterion, that is, under certain linear constraints, the signal power can be minimized. Based on the LCMV criterion, the response vector in the constraint direction is constant 1, while the constraint matrix is the desired signal direction vector,

$$\begin{cases} s.t. W^H a(\theta_p) = 1 \\ \min P_{out} = E\{|y(n)|^2\} \end{cases} \tag{14}$$

MVDR algorithm optimal weight vector is calculated as

$$W_{MVDR} = R_{XX}^{-1} a(\theta_p)\left(\alpha^H(\theta_p) R_{XX}^{-1} a(\theta_p)\right)^{-1} \tag{15}$$

IOP Conf. Series: Earth and Environmental Science **114** (2018) 012010 doi:10.1088/1755-1315/114/1/012010

The signal after interference mitigation is

$$S_{MVDR}(n) = W_{MVDR}^{H} a(\theta_p)s(n) \quad (16)$$

4. Simulation results

With Matlab simulation, the basic parameters are set: RF frequency f = 1575.42 MHz, the signal-to-noise ratio of the signal arriving at the antenna segment is about -28 dB, the spreading code rate is 1.023 MHz, the sampling rate is 1 MHz, the interference intensity is obviously stronger than the noise, The Jamming Signal Rate (JSR) is set in the range of 30 dB to 80 dB depending on the scene. In the case of most satellite navigation interference environments, the array model uses a typical 8-element linear array, Spacing d is 1/2 RF signal wavelength, interference for broadband Gaussian interference, bandwidth of 1.023 MHz, the center frequency and signal frequency of the same, as shown in Figure 3. The two algorithms pour into the hardware, the Central Business District in Beijing for the actual road test, we set the sampling rate of 1hz, positioning results shown in Figure 4. The yellow line represents the use of receiving RTK (Real - time kinematic) high precision , the red line represents the position data obtained using the MVDR algorithm with multi-antenna arrays, and the green line represents the position data obtained using the PI algorithm with multi-antenna arrays.

Figure 4 and Figure 5 show that the MVDR algorithm has a certain advantage over the RTK, but it has more obvious advantages than the PI algorithm. The anti-multipath interference ability is obvious and the contribution to the positioning accuracy is greater.

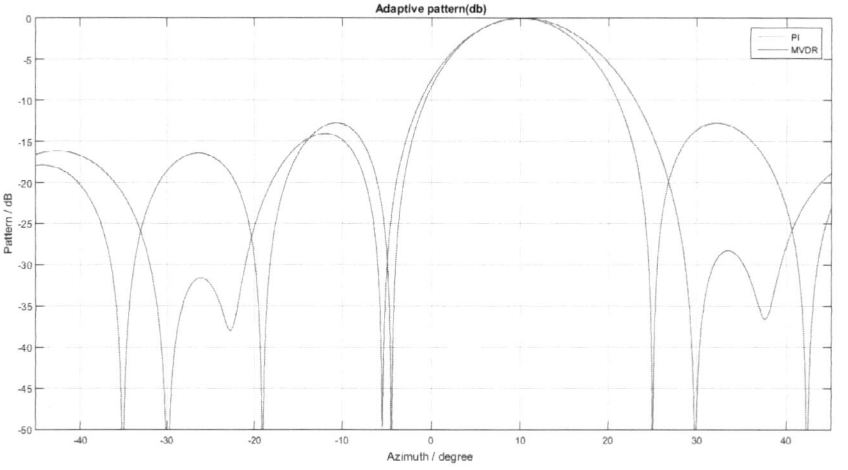

Figure 3. Adaptive Pattern

Figure 4. Comparison positioning data between PI and MVDR in vehicle test

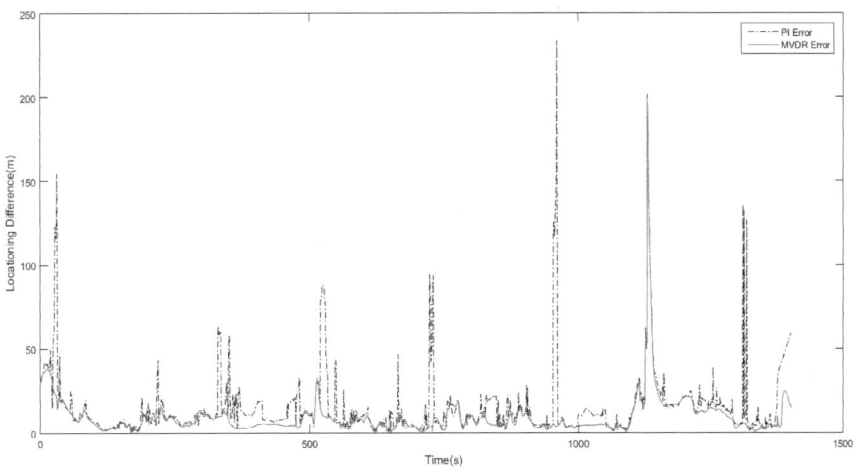

Figure 5 Comparison Error between PI and MVDR in the vehicle test

Figure 4 and Figure 5 show that the MVDR algorithm has a certain advantage over the RTK, but it has more obvious advantages than the PI algorithm. The anti-multipath interference ability is obvious and the contribution to the positioning accuracy is greater.

5. Conclusion

In this paper, two algorithms are proposed based on adaptive antenna array, the simulation results show that the MVDR algorithm is better than the classical PI algorithm in the application of the vehicle terminal, and it is better than the multi - path interference. The validity of the application of MVDR algorithm in vehicle high precision receiver is verified by on-the-road test, which lays the foundation for the application of multi-antenna array antenna in the vehicle.

6. References

[1] Groves P,2013 Multipath vs. NLOS signals. How Does Non-Line-of-Sight Reception Differ From Multipath Interference *J.GNSS Solutions* **8** pp 40-42

[2] Daniel Egea, Jos´ eA.L´opez-Salcedo and Gonzalo Seco-Granados, 2014 Interference and multipath sequential tests for signal integrity in multi-antenna GNSS receivers *2014 Proc. of the IEEE Sensor Array and Multichannel Signal Processing Workshop* vol 14 pp 117-120

[3] A. Italiano, F. Principe, R. Cioni and R. Perago,2010 Multipath and interference modelling in complex GNSS scenarios *Proc. of the Fourth European Conference on Antennas and Propagation* pp 1-5

[4] Wenting Yi, Weiwei Song, Yidong Lou, Chuang Shi & Yibin Yao, 2016 A method of undifferenced ambiguity resolution for GPS+GLONASS precise point positioning, *Scientific Reports vol* 6 pp 1-13

[5] E. D. Kaplan and C. J.Hegarty, 2006 *Understanding GPS—Principles and Applications* (Norwood, Artech House)

[6] Marathe T, Daneshmand S, Lachapelle G,2016 Assessment of measurement distortions in GNSS antenna array space-time processing. *J.International Journal of Antennas and Propagation* vol 2 pp 1-17

[7] Huang Tao, Chang Qing, Xu Yong, 2016 Comparative analysis of two kinds of anti-jamming algorithms based on LCMV *J. Journal of Terahertz Science and Electronic Information Technology* vol 14 pp 891-896

[8] Vagle, N, Broumandan, A., Jafarnia-Jahromi, A. et al. 2016 Performance analysis of GNSS multipath mitigation using antenna arrays *J.Global Position. Syst.*.vol 14 p 4

IOP Conf. Series: Earth and Environmental Science **114** (2018) 012011 doi:10.1088/1755-1315/114/1/012011

Power Allocation and Outage Probability Analysis for SDN-based Radio Access Networks

Yongxu Zhao, Yueyun Chen and Zhiyuan Mai

School of Computer and Communication Engineering, University of Science and Technology Beijing, Beijing, China

E-mail: zhao_yx@yeah.net; chenyy@ustb.edu.cn; mcqueensee@outlook.com

Abstract. In this paper, performance of Access network Architecture based SDN (Software Defined Network) is analyzed with respect to the power allocation issue. A power allocation scheme PSO-PA (Particle Swarm Optimization-power allocation) algorithm is proposed, the proposed scheme is subjected to constant total power with the objective of minimizing system outage probability. The entire access network resource configuration is controlled by the SDN controller, then it sends the optimized power distribution factor to the base station source node (SN) and the relay node (RN). Simulation results show that the proposed scheme reduces the system outage probability at a low complexity.

1. Introduction

In recent years, with the popularity of personal mobile smart terminal devices and smart phones, mobile data traffic has shown a rapid growth trend. Cisco's 11th Annual Mobile Visual Networking Index (VNI) report predicts that global mobile data traffic will grow sevenfold between 2016 and 2021[1]. The main challenge for the fifth generation mobile communication (5G) network will be to effectively expand the range of wireless communications, improve the intelligent level of the system[2] and increase the transmission rate of users at the edge of the cell to meet diverse service requirements. In order to meet this challenge, we need to propose a new access network architecture, in this paper, SDN will be introduced into the wireless access network to improve the intelligent level of the relay access network.

SDN separates the control plane from the data plane as a new network architecture, it treats the network as a complete operating system and extracts the application from the underlying hardware device so that management-related functions are centralized in the controller[3] , [4]. Therefore, the intelligence of the network will be achieved through a centralized controller to manage the global network[5]. In the SDN-based access network architecture, the control plane and data plane separation, control plane through the SDN controller to achieve the entire access network resource allocation of centralized management, the SN and the RN obtains the optimal power allocation factor through the PSO-PA algorithm calculated by the SDN controller[6], finally the minimized system outage probability is obtained and the quality of service of the system is improved. The interactive protocol between the SDN controller and the data plane draws on the OpenFlow protocol[7].

At present, more and more scholars have started to study the introduction of SDN technology into mobile cellular networks. In[8], a load balancing network under SDN architecture was proposed. Load balancing was implemented by a load balancing algorithm based on utility function. The network state and user benefit were taken into account at the same time. However, when the user application request is large, Function calculation is more complex, less efficient. In[9] ,a SDN-based SoftCell architecture was presented. SoftCell is an SDN-based cellular core network that guided the user's data traffic through an intermediate device link, consolidated valid forwarding rules, reduced the forwarding flow of data traffic, and when the packet entered the switch, it was classified to improve the flexibility and scalability of the cellular core network, but the architecture does not analyze the application of SDN in the access network. In[10], an optimization algorithm of combined power distribution and relay location was proposed. Under the condition of total power, the optimal solution was obtained by solving the equation, and the algorithm was high in complexity.

Based on the above problem, this paper studies the resource allocation algorithm of relay access network based on SDN architecture, and proposes an efficient power allocation algorithm (PSO-PA), which can get the optimal in the case of fewer iterations, the power allocation factor is sent by the SDN controller to the SN and the RN through the feedback message in the OpenFlow protocol, minimizes the system fault probability, and improves the service quality of the system.

2. System model

Introducing SDN technology into the access network will make the access network more flexible, efficient, intelligent and open. SDN-based heterogeneous access network system model, shown in Figure 1, is a multi-level architecture including macro cells, microcells, and relays. A set of macro cell base stations is controlled by a centralized SDN controller which connects to base stations through high-capacity fibers, controls the data plane according to OpenFlow protocol. In order to reduce the pressure on the SDN controller, the controller only associates with the set of macrocell base stations connected to it. Relays periodic report their load conditions to the macrocell via a wireless or fiber backhaul link.

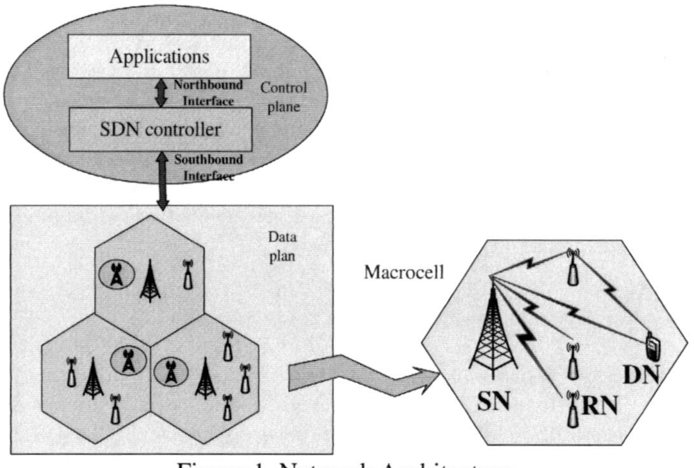

Figure 1. Network Architecture

SDN controller can manage the resource of the whole wireless access system including spectrum allocation, power allocation, access control and switching control. And the power controller can control the nodes of the access network through the OpenFlow protocol.

The interactive protocol between the power splitter in SDN controller and the access network draws on the OpenFlow protocol between the controller and the switch in the SDN architecture. The control information, including access management, switching control and other messages, sent by the power divider to the wireless access point, is realized by configuration messages in the OpenFlow message. Inquiring message is implemented through read-state messages in the Openflow protocol. The SDN controller can obtain the access status of each access station by inquiring to schedule the radio resource more effectively. The feedback message utilizes the packet-out message in the Openflow protocol, which is the optimal power allocation factors calculated by the power splitter algorithm described in this paper.

As is shown in Figure 1, SN, RN, DN represent the Base station Source Node, Relay Node and Destination Node respectively. The relaying strategy is decode-forward and whole communication process is divided into two stages. In the first stage, SN sends the information x. RN and the DN receive x at the same time. In the second stage, SN does not send any message while RN send $\hat{x}$ to DN, $\hat{x}$ is the re-encoded signal. DN combines the signals received in two time slots according to the maximum ratio.

In the first stage, the SN sends the signal to the RN and DN. Respectively, the signals received by the RN and the DN are

$$Y_{s,r} = \sqrt{P_0} H_{s,r} x + N_{s,r} \tag{1}$$

$$Y_{s,d} = \sqrt{P_0} H_{s,d} x + N_{s,d} \tag{2}$$

Where, P_0 denotes the SN transmission power. $H_{s,r}$ denotes the Rayleigh channel gain between SN and RN, $H_{s,d}$ denotes the Rayleigh channel gain between SN and DN, The mean value of the Rayleigh distribution is 0, and the variance σ_n^2 follows the complex Gaussian distribution. $N_{s,r}$ denotes additive white Gaussian noise (AWGN) between SN and RN, $N_{s,d}$ denotes additive white Gaussian noise (AWGN) between SN and DN. The mean value and the power spectral density of AWGN is 0 and σ_n^2.

In the second stage, the RN will re-encode the signal of the received base station node and send it to the DN. The received signal can be expressed as follows

$$Y_{r,d} = \sqrt{P_r} H_{r,d} \hat{x} + N_{r,d} \tag{3}$$

Where P_r represents the relay transmission power. $H_{r,d}$ represents the Rayleigh channel gain between RN and DN. The mean of the Rayleigh distribution is 0, and the variance $\sigma_{r,d}^2$ follows the complex Gaussian distribution. $\hat{x}$ is the signal sent from the SN and processed by the relay, the DN combines the signals received in two stage according to the maximum ratio. The combined signal can be expressed as

$$Y_d = b_1 Y_{s,d} + b_2 Y_{r,d} \tag{4}$$

$$b_1 = \frac{1}{N_0} \sqrt{P_0} H_{s,d}^* \tag{5}$$

$$b_2 = \frac{1}{N_0} \sqrt{P_r} H_{r,d}^* \tag{6}$$

Where, $H_{s,d}^*$, $H_{r,d}^*$ are the conjugate complex numbers of $H_{s,d}$ and $H_{r,d}$ respectively. N_0 represents the noise power.

3. Outage probability analysis

3.1. Relay selection

For the time-varying wireless communication channel, when the communication system of the channel mutual information I is greater than the system requires the minimum transmission rate R, the system can work properly. Otherwise, when $I < R$,the system will not be able to communicate properly, determine the outage[11]. In the first stage, the SN sends a signal to the RN, and the mutual information between the SN and the RN is

$$I_{s,r} = \frac{1}{M+1} \log_2 \left(1 + r_{s,r}\right) \tag{7}$$

$$r_{s,r} = P_0 H_{s,r} / N_0 \tag{8}$$

Where, $M \in \{0,1\}$,when relay is involved in communication $M=1$, otherwise, $M=0$. $r_{s,r}$ is the instantaneous signal-noise ratio(SNR) of the SN to the RN, $H_{s,r}$ is Channel gain, subject to an exponential distribution with a parameter of $1/\sigma^2$.
If the channel quality between the SN and the RN in high condition, the actual transmission rate of the system is greater than the minimum transmission rate required by the system. The system can support the data transmission of the target transmission rate R ; but when the mutual information between the two nodes is less than the target transmission rate R , the system is outage and the system outage probability can be expressed as

$$P_{out} = P\left\{I = \frac{1}{2}\log_2(1+r) < R\right\} = P\left\{r < 2^{2R}-1\right\} \tag{9}$$

$$P_{out} = \int_0^{2^{2R}-1} \frac{1}{\sigma^2}\exp\left(-\frac{r}{\sigma^2}\right)dr = 1-\exp\left(-\frac{2^{2R}-1}{\sigma^2}\right) \tag{10}$$

The set of Φ is the nodes of N relays that can successfully decode transmit the signal. The set Φ satisfies the instantaneous SNR of the base station to relay not less than the system requirement threshold γ_{th}

$$\Phi \in \left\{ i \mid \left(P_0 |H_{s,r}|^2 / N_0\right) \ge \gamma_{th} \right\} \tag{11}$$

According to the probability distribution density, the probability of Φ is

$$P\{\Phi\} = \prod_{i\in\Phi} P(r_{s,r} \ge \gamma_{th1}) \prod_{j\notin\Phi} P(r_{s,r} \le \gamma_{th1}) = \prod_{i\in\Phi}\exp(-\gamma_{th1}/\sigma_{s,r}^2) \prod_{j\notin\Phi}\left[1-\exp(-\gamma_{th1}/\sigma_{s,r}^2)\right] \tag{12}$$

where, $\gamma_{th1} = \dfrac{2^{2R}-1}{P_0 / N_0}$.

In the second stage, in order to maximize the received SNR at the DN, we select the largest instantaneous SNR relay for signal forwarding, where, $r_{opt} = \arg\max_{i\in\Phi} r_{i,d}$.

3.2. Power allocation

The total power of SN and RN is P, let $p_0 = \alpha p$, $p_r = (1-\alpha)p$, α represents the power allocation factor, The SNR at the DN can be expressed as

$$SNR_{i,d} = \alpha\lambda_{s,d}\frac{P}{N_0} + (1-\alpha)\lambda_{r,d}\frac{P}{N_0} \tag{13}$$

Assume the transmission link that SN to DN is link T_1, the transmission link SN to RN is T_2, and the transmission link RN to DN is T_3.If the system is outage, Then the T_1 is outage and there is a link outage in T_2 and T_3.
The outage probability of the transmission link from the source node to the DN can be expressed as

$$P_{s,d} = P\left\{\frac{1}{2}\log_2(1+\alpha\lambda_{s,d}\frac{P}{N_0}) \le R\right\} = P\left\{\alpha\lambda_{s,d}\frac{P}{N_0} < 2^{2R}-1\right\} = 1-\exp\left[-\frac{\left(2^{2R}-1\right)}{\sigma_{s,d}^2\alpha P/N_0}\right] \tag{14}$$

The outage probability of the transmission link of the SN to the RN can be expressed as

$$P_{s,r} = P\left\{\frac{1}{2}\log_2(1+\max_{i\in\Phi}\alpha\lambda_{s,r}\frac{P}{N_0}) \le R\right\} = P\left\{\max_{i\in\Phi}\alpha\lambda_{s,r}\frac{P}{N_0} < 2^{2R}-1\right\} = \prod_{i=1}^{N} 1-\exp\left[-\frac{\left(2^{2R}-1\right)}{\sigma_{s,r}^2\alpha P/N_0}\right] \tag{15}$$

The conditional outage probability that the transmission link from the RN to the DN can be expressed as

$$P\{outage|\Phi\}=P\left\{\frac{1}{2}\log_2\left(1+\max_{i\in\Phi}SNR_{r,d}\right)\le R\right\}=\prod_{i\in\Phi}P\left\{\frac{1}{2}\log_2\left(1+SNR_{r,d}\right)\le R\right\}$$
$$=\prod_{i\in\Phi}P\left\{\frac{1}{2}\log_2\left(1+(1-\alpha)\lambda_{r,d}\frac{P}{N_0}\right)\le R\right\}=1-\exp\left[-\frac{\left(2^{2R}-1\right)}{\sigma_{s,d}^2(1-\alpha)P/N_0}\right] \tag{16}$$

The probability that the RN to the DN link is outage under the condition that the link between the SN and the RN can work properly can be expressed as

$$P_{r,d}\{outage\}=\sum_{i=1}^{N}\sum_{\Phi}P\{outage|\Phi\}P\{\Phi\}$$
$$=\sum_{i=1}^{N}\sum_{\Phi}1-\exp\left[-\frac{\left(2^{2R}-1\right)}{\sigma_{s,d}^2(1-\alpha)P/N_0}\right]\times\prod_{i\in\Phi}\exp\left(-\frac{\left(2^{2R}-1\right)}{\sigma_{s,d}^2\alpha P/N_0}\right)\prod_{j\notin\Phi}\left[1-\exp\left(-\frac{\left(2^{2R}-1\right)}{\sigma_{s,d}^2\alpha P/N_0}\right)\right] \tag{17}$$

According to the outage probability of each link, the total outage probability of the system is

$$P_{out}=\left\{1-\exp\left[-\frac{\left(2^{2R}-1\right)}{\sigma_{s,d}^2\alpha P/N_0}\right]\right\}\left\{\prod_{i=1}^{N}1-\exp\left[-\frac{\left(2^{2R}-1\right)}{\sigma_{s,r}^2\alpha P/N_0}\right]+\sum_{i=1}^{N}\sum_{\Phi}1-\exp\left[-\frac{\left(2^{2R}-1\right)}{\sigma_{s,d}^2(1-\alpha)P/N_0}\right]\right.$$
$$\left.\times\prod_{i\in\Phi}\exp\left(-\frac{\left(2^{2R}-1\right)}{\sigma_{s,d}^2\alpha P/N_0}\right)\prod_{j\notin\Phi}\left[1-\exp\left(-\frac{\left(2^{2R}-1\right)}{\sigma_{s,d}^2\alpha P/N_0}\right)\right]\right\} \tag{18}$$

3.3. Power allocation of PSO

Considering the limitation and complication of traditional Iterative methods to obtain the optimal solution, we propose a PSO-PA algorithm to optimize the non-linear outage probability formula. The PSO-PA algorithm is designed to introduce particle swarm algorithm[12] into power allocation and obtain the optimal power distribution factor under the minimized outage probability. The PSO-PA algorithm we proposed is a global optimization algorithm based on the behavior of the bird population. Besides, PSO-PA algorithm has a great advantage, such as easy to implement, less parameters, the faster convergence velocity and accuracy and so on.

3.3.1. Initialization. Assume the target search space dimension D, and each particle represents an interior point in the space to be searched. The population size is M and contains M particles is $X=(x_1,\cdots,x_i,\cdots,x_D)$. $X_i=(x_{i1},\cdots,x_{i2},\cdots,x_{iD})^T$ and $V_i=(v_{i1},\cdots,v_{id},\cdots,x_{iD})^T$ represent the position and the velocity of the i-th particle respectively.

3.3.2. Search and optimization. The optimal position searched in i-th particle is local optimum, denoted as $p_i=(p_{i1},\cdots,p_{id},\cdots,p_{iD})^T$. And the optimal position searched in whole population is global extremum, denoted as $p_g=(p_{g1},p_{g2},\cdots p_{gD})^T$. Particles are kept close to the optimal position of the current population. The velocity and position of the particles will be updated following as:

$$v_{id}(t+1)=wv_{id}(t)+c_1r_1(t)\left(p_{id}(t)-x_{id}(t)\right)+c_2r_2(t)\left(p_{gd}(t)-x_{id}(t)\right) \tag{19}$$

$$x_{id}\left(t+1\right)=x_{id}\left(t\right)+v_{id}\left(t+1\right) \tag{20}$$

where $d=1,2,\cdots,D$, $i=1,2,\cdots,M$. M represents population size. t is the number of iterations. w is Inertia factor. r_1 and r_2 are random numbers between 0 and 1, and they are used to keep the diversity of population. c_1 and c_1 are learning factors to made the particles self-learning, so that the particles could constantly adjust their direction based on the historical optimal location of particles and population.

3.3.3. Matching by didirection. Individual optimum is updated continually in the iterative process of particle swarm, then searching the optimal power distribution factor, and the minimized outage probability will be obtained.

Based on the above PSO algorithm and the non-linear outage probability formula, we propose PSO-PA algorithm that think of power allocation factor as the particle in the whole population. We can obtain the optimal particle position using the proposed algorithm, that is, it's effective to obtain the optimal power allocation factor under the minimized outage probability.

4. Simulation results

In the simulation model, the system parameters are set as follows: the number of relay in the system is N; the minimum transmission rate is R=1bit/s/Hz under the condition of non-interrupting; $\sigma^2_{s,r}$ and $\sigma^2_{s,d}$. are the channel gain variance from SN to RN and RN to DN that both satisfy the random distribution of [0,2]; the channel gain variance $\sigma^2_{s,d}$ of the SN to the DN is 0.7.In the PSO-PA algorithm, the population size sizepop=20,the maximum number of iterations is 100,the learning factors c_1 and c_2 are both 0.4, the maximum value of the w is 0.8 and the minimum value is 0.2.

In Figure 2, the performance of the proposed PSO-PA power allocation algorithm and Uniform power allocation algorithm are compared and analyzed by simulation, the outage probability between different algorithms is given under different SNR. It can be seen from the simulation diagram that with the increase of SNR, the outage probability under all algorithms is reduced. The intelligent algorithm that using the PSO-PA power allocation algorithm expressed more obvious performance improvement about the outage probability. In the case of the same SNR, the performance of the proposed PSO-PA power allocation algorithm is better than that of the Uniform power allocation algorithm, which verifies the effectiveness of the proposed algorithm.

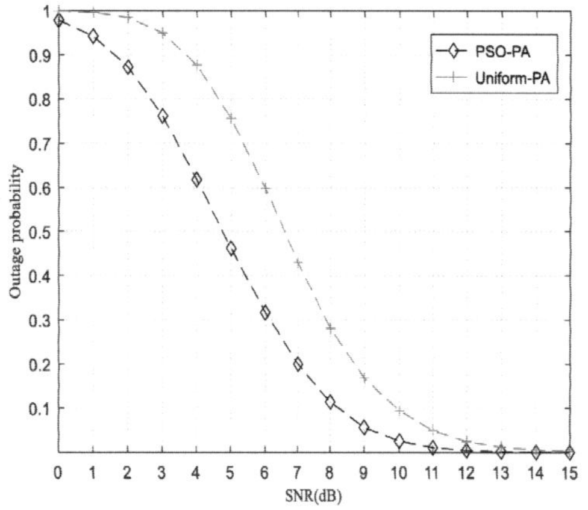

Figure 2. Outage probability of PSO-PA and UPA

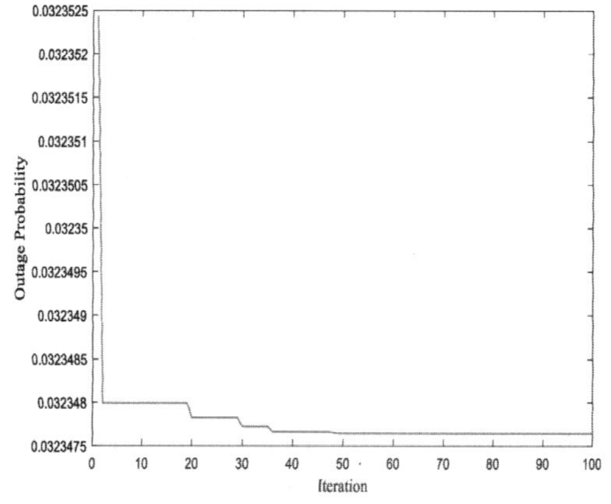

Figure 3. The convergence performance

Figure 3 shows us that the proposed PSO-PA algorithm in the 35th iteration can find the optimal solution, the algorithm has the advantage of fast iteration, can effectively improve the access network resource allocation ability.

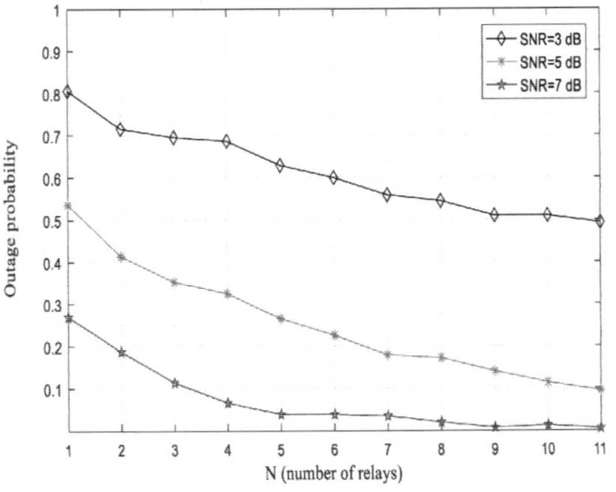

Figure 4. The system outage performance under different SNR

In Figure 4, we can see the comparison of the outage performance about the proposed PSO-PA power allocation algorithm in different SNR. When the SNR is less than 5 dB, the outage probability of the system is at a high level, while with the increase of SNR, the system's outage probability is greatly reduced, so the SNR of the system have a great impact on the outage probability. When the number of relays increases to a certain number, the effect of the relays' increasing on the outage probability of the system is reduced. Therefore, after adding the relay numbers to a certain value, it is necessary to consider increasing the SNR of the system to further reduce the system's outage probability.

5. Conclusion

In this paper, we formulate the relay selection and power allocation problem in the radio access network based on SDN. To improve the outage probability performance of the system, the optimization of the power allocation is investigated. We propose a scheme called PSO-PA based on the Particle Swarm Optimization algorithm to obtain the optimal power allocation factor. Finally, the simulation results show that the proposed PSO-PA algorithm is effective. It enhances the system performance significantly in outage probability at a low complexity.

6. References

[1] Index V N. Cisco Visual Networking Index: Global Mobile Data Traffic Forecast Update, 2016–2021 White Paper[J].J.P. Secilla and N. Garcia,"Template location in noisy pictures,"Signal Process. **14**, 347-361, (1987).

[2] Demestichas P, Georgakopoulos A, Tsagkaris K, et al. Intelligent 5G Networks: Managing 5G Wireless/Mobile Broadband[J]. *Vehicular Technology Magazine IEEE*, 2015, 10(3): 41-50A. Rosenfeld and A. Kak. *Digital Image Processing* (2nd Edition, Vol. 2 ed.),Academic Press, Orlando (1982).

[3] Ali-Ahmad H, Cicconetti C, Oliva A D L, et al. An SDN-Based Network Architecture for Extremely Dense Wireless Networks[C]. *Future Networks and Services*, 2013: 1-7.

[4] Wang K, Wang Y, Zeng D, et al. An SDN-Based Architecture for Next-Generation Wireless Networks[J]. *IEEE Wireless Communications*, 2017, 24(1): 25-31.

[5] Sun S, Gong L, Rong B, et al. An intelligent SDN framework for 5G heterogeneous networks[J]. *Communications Magazine IEEE*, 2015, 53(11): 142-147.

[6] Nguyen-Duc T, Kamioka E. An extended SDN controller for handover in heterogeneous wireless network[C]. *The Asia- Pacific Conference on Communications*, 2015.

[7] Martinez C, Ferro R, Ruiz W. Next generation networks under the SDN and OpenFlow protocol architecture[C]. *Engineering Applications - International Congress on Engineering*, 2015: 1-7.

[8] Tang W, Liao Q. An SDN-Based Approach for Load Balance in Heterogeneous Radio Access Networks[C]. *Computer Applications and Communications*, 2014: 105-108.

[9] Jin X, Li L E, Vanbever L, et al. SoftCell:scalable and flexible cellular core network architecture[C]. *ACM Conference on Emerging NETWORKING Experiments and Technologies*, 2013: 163-174.

[10] Shen Q, Juntti M, Matsumoto T. Joint Optimization of Power Allocation and Relay Position for Lossy-Forwarding Relaying[C]. *IEEE Wireless Communications and NETWORKING Conference*, 2016.

[11] Ozarow L H, Shamai S, Wyner A D. Information theoretic considerations for cellular mobile radio[J]. *Vehicular Technology IEEE Transactions on*, 1994, 43(2): 359-378.

[12] Trelea I C. The particle swarm optimization algorithm: convergence analysis and parameter selection[J]. *Information Processing Letters*, 2003, 85(6): 317-325.

Design of c-band telecontrol transmitter local oscillator for UAV data link

Hui Cao[1] Yu Qu[1] and Zuxun Song[1]
1 Institute 365, Northwestern Polytechnical University, 127 YouYi Xilu Xi'an 710072 Shaanxi, P.R.China
E-mail: nash790127@nwpu.edu.cn and 6290333@qq.com

Abstract. A C-band local oscillator of an Unmanned Aerial Vehicle (UAV) data link radio frequency (RF) transmitter unit with high-stability, high-precision and lightweight was designed in this paper. Based on the highly integrated broadband phase-locked loop (PLL) chip HMC834LP6GE, the system performed fractional-N control by internal modules programming to achieve low phase noise and small frequency resolution. The simulation and testing methods were combined to optimize and select the loop filter parameters to ensure the high precision and stability of the frequency synthesis output. The theoretical analysis and engineering prototype measurement results showed that the local oscillator had stable output frequency, accurate frequency step, high spurious suppression and low phase noise, and met the design requirements. The proposed design idea and research method have theoretical guiding significance for engineering practice.

1. Introduction

Frequency source is a module that generates a large number of discrete frequencies with the same stability and accuracy as the reference signal through linear operations in the frequency domain by one or more reference sources with high frequency stability and accuracy. It is an important front part of RF communication system. With the development of UAV data link system, the frequency source has also been put forward higher and higher requirements. This is mainly reflected in: fast conversion speed, wide frequency band coverage, small frequency resolution, high stability, low spurious, low phase noise, small size, light weight and multi-function [1][2].

The HMC834LP6GE is a low noise, wide band, fractional-N PLL chip that features an integrated Voltage Controlled Oscillator (VCO) with a fundamental frequency of 2800 MHz - 4200 MHz, and an integrated VCO Output Divider (divide by 1/2/4/6.../60/62) and doubler, that together allow the HMC834LP6GE to generate frequencies from 45 MHz to 1050 MHz, from 1400 MHz to 2100 MHz, from 2800 MHz to 4200 MHz, and from 5600 MHz to 8400 MHz. The integrated Phase Detector (PD) and delta-sigma modulator, capable of operating at up to 100 MHz, permit wider loop-bandwidths with excellent spectral performance. The HMC834LP6GE features industry leading phase noise and spurious performance, across all frequencies, that enable it to minimize blocker effects, and improve receiver sensitivity and transmitter spectral purity [3][4].

High-performance frequency source of UAV data link system based on HMC834LP6GE can be designed by adding simple peripherals, such as power supply circuit, reference circuit and loop filter.

2. System design

The indicators required by an UAV telecontrol transmitter local oscillator are: The output frequency is 5600 ~ 5800MHz, with 8MHz step; the frequency stability is better than 3ppm; the phase noise is better than -80dBc / Hz / 10 KHz; the size of the circuit board is controlled within 120mm × 60mm.

According to the system requirements, the local oscillator of the telecontrol transmitter based on the integrated PLL chip HMC834LP6GE was designed; the block diagram is shown in figure 1. PD, CP (charge pump) and VCO are integrated in the chip HMC834LP6GE. 100MHz temperature compensated crystal oscillator (TCXO) was chosen as the reference source (REF). Considering the phase noise and the frequency interval, we selected 100MHz as the PD frequency, and made the PLL chip operating under the fractional-N mode. The low pass filter (LPF) was designed as a 4-stage passive filter and a 2-stage amplifier was adopted. Power supply consisted of three + 5V and a + 3.3V. The AVR chip ATmega8L was selected as the microcontroller (MCU).

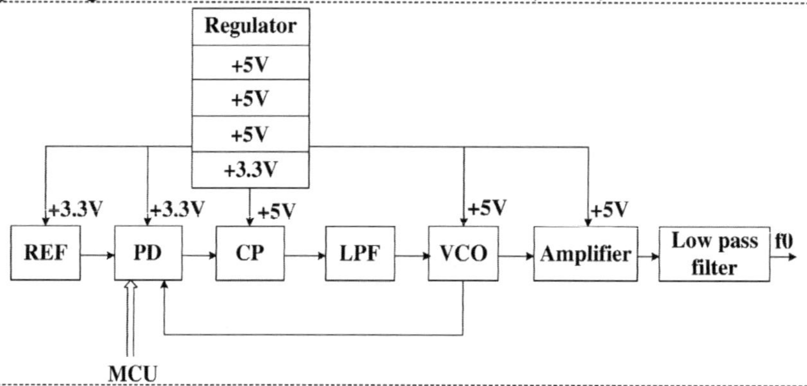

Figure 1. Block diagram of the system.

3. Hardware design

3.1. Reference circuit design

An excellent reference source is important for a frequency source system. Reference source performance is good, and then the choice of synthesis and device will have a larger space [5]. High-performance 100MHz TCXO WTX-503CE57C3.3-100.000MHz was chosen according to performance, cost and volume. The size of the TCXO is 5 * 3.2mm, with 3.3 V power supply, and the current is 10mA. The output waveform is HCMOS and the frequency stability is 0.5ppm. The TCXO has the advantages of high output frequency, high stability, small size and low power consumption, which can satisfy the design requirements. The reference frequency source circuit is shown in figure 2. The reference input pin of synthesizer was connected with a 100Ω resistor which connected to the ground in order to ensure 50Ω matching; an AC coupling capacitor was connected also.

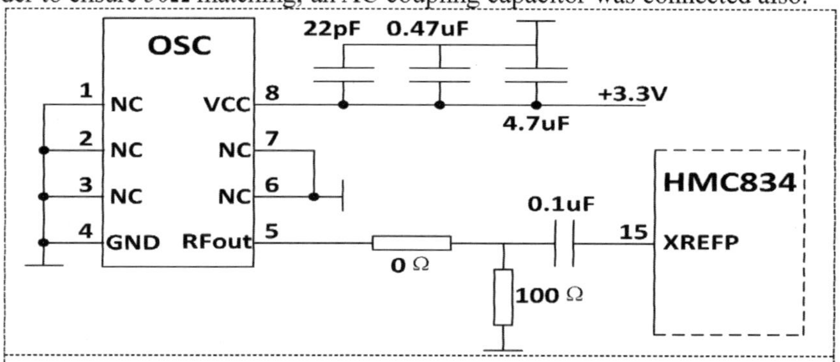

Figure 2. Circuit of the reference frequency source.

3.2. Power supply design

There are a lot of power pins in HMC834 chip. In order to achieve a better anti-interference performance of the circuit, the method of separate power supply was adopted. VCC1 and VCC2 of the chip shared a single 5V linear regulator, which avoid the VCO affecting by interferences. CP power supply pins VPPCP and VDDLS shared a 5V linear regulator. The 2-stage power amplifiers used a 5V linear regulator. The rest power supply pins, MCU and TCXO shared a 3.3V linear regulator together. The voltage regulators adopted two models: LT1763CS8 with +5V output voltage and LT1963AEST with +3.3V output voltage.

3.3. Loop filter design

The loop filter is located between CP and VCO, it not only filters out the noise from the crystal, the output noise of the PD itself and the carrier frequency , as well as the leakage of the PD frequency, but also filters out the noise from the VCO [6]. The design of loop filter is the key of the design of a frequency synthesizer, and the technical indicators of the system should be rational considered. The bandwidth of the filter design needs to be compromised, if the bandwidth is narrow, the noise influence from VCO is serious and the loop locking time is extended; if the bandwidth is wide, the noise influence from crystal oscillator and PD is serious [7]. In this design, the TCXO with better phase noise was selected as the reference source, which makes the near-end noise relatively small. For the noise of the far-end, the noise from VCO in HMC834 plays a leading role. Therefore, in order to suppress the high frequency noise of VCO, the loop bandwidth should be set wider [8][9].

ADI Company's ADIsimPLL4.2 software was adopted in this design. PLL circuit can be designed more convenient by using the software through the following operations: chip selection, frequency divider selection, PD frequency selection, reference frequency selection and phase noise model input, loop bandwidth, phase margin setting, loop filter selection, etc. [10]

Three loop bandwidth of 125KHz, 75KHz and 18.9KHz were simulated in this design, four-order passive low pass filters were adopted in all the three designs. Figures 3(a), 3(b) and 3(c) are the phase noise curves of frequency 5700MHz of the three kinds of bandwidth, and table 1 is the comparison of the phase noises. It can be seen that the low frequency offset phase noise is poor when the loop bandwidth is very narrow, while the high frequency offset phase noise behaves fairly well under all the three bandwidths. According to the comparison, the best phase noise performance was the bandwidth of 125 KHz, so the final selection of bandwidth was 125KHz. The circuit is shown in figure 4. The ideal value of phase noise is -107.1dBc/Hz/10KHz, which obviously meets the design requirements.

Table 1. Phase noise table.

Frequency	Phase Noise		
	125KHz	75KHz	18.9KHz
100Hz	-92.83	-92.83	-91.93
1.00KHz	-102.3	-101.9	-88.20
10.0KHz	-107.1	-101.2	-79.69
100KHz	-102.2	-101.4	-104.9
1.00MHz	-130.7	-131.1	-131.3

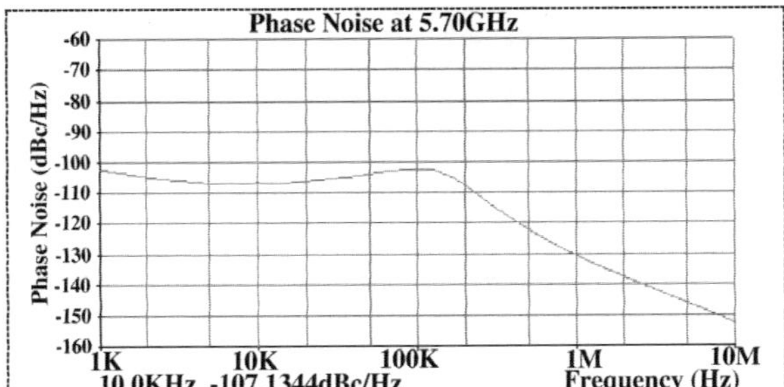

(a)

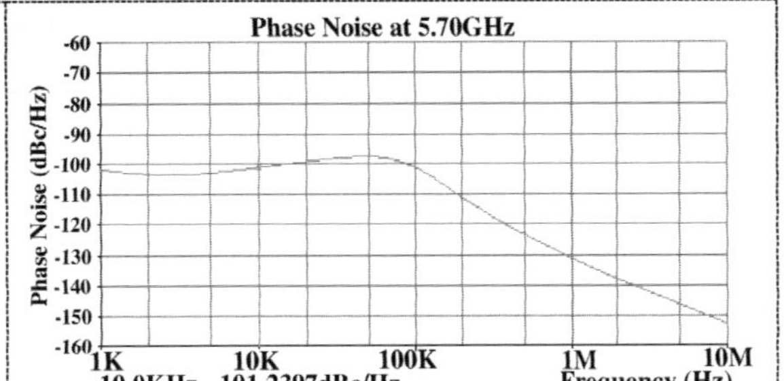

(b)

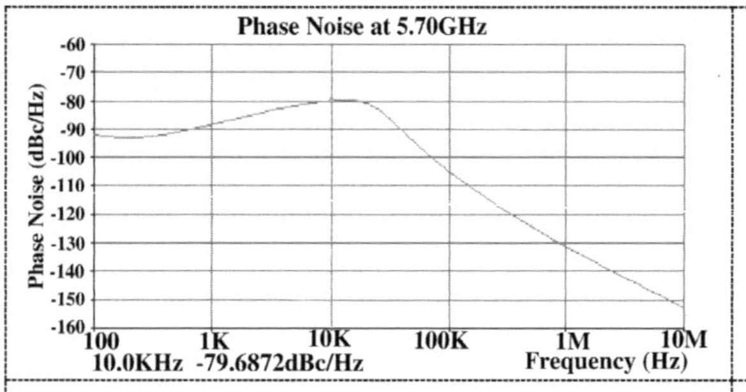

(c)

Figure 3. Simulation results of phase noise curves at different loop bandwidth, (a) bandwidth=125KHz, (b) bandwidth=75KHz, (c) bandwidth=18.9KHz.

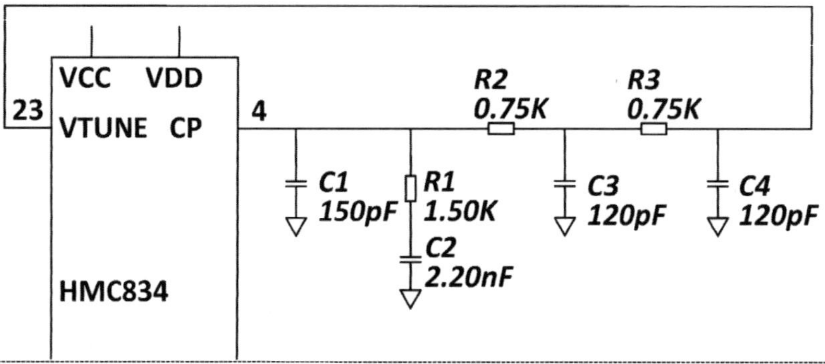

Figure 4. The parameters of loop bandwidth 125KHz.

4. Software design

4.1. Registers configuration

There are a total of 18 PLL registers and 7 VCO subsystem registers in HMC834. PLL registers are from Reg00h to Reg13h, including multiple frequency division coefficient registers, charge pump register, lock detect register, etc. VCO subsystem registers are from VCO_Reg00h to VCO_Reg06h, including VCO frequency division coefficient setting, the output power setting and the signal output setting, etc. The operation of the VCO subsystem registers is achieved by writing the PLL register Reg05h [11]. The default frequency of 5700MHz is configured in the following order shown in table 2.

Table 2. Register configuration.

Order No.	Registers	Value
1	Reg01h	0x000002
2	Reg02h	0x000001
3	Reg03h	0x00001C
4	Reg05h	0x007E28
5	Reg05h	0x00B6A0
6	Reg05h	0x002018
7	Reg05h	0x000000
8	Reg06h	0x030F4A
9	Reg07h	0x00014D
10	Reg08h	0xC1BEFF
11	Reg09h	0x547FFF
12	Reg0Ah	0x002046

Order No.	Registers	Value
13	Reg0Bh	0x07C021
14	Reg0Ch	0x000000
15	Reg0Fh	0x000001
16	Reg04h	0x800000

4.2. Software implementation

Only writing operation is needed when configuring the HMC834 registers in this design. When communicating, the MCU ATmega8L operates as the master device, HMC834 as the slave device. The master and slave devices communicate via the three-wire serial interface SEN (serial enable port), SDI (serial data input) and SCK (serial clock). When SEN is pulled high, communication is effective. SCK as clock trigger signal is generated by the delay program of the MCU, and the whole communication needs 32 clock cycles. The first data is the judgment of reading and writing, when the first clock arrives, the data bit is low to indicate the write command, the following 6 clocks are used to determine the register address, and the next 24 clocks for data transmission. When the SEN cleared, the write cycle is completed. The system software flow chart is shown in figure 5.

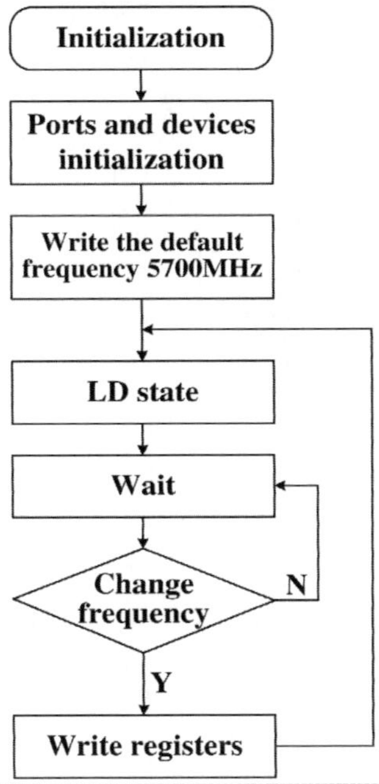

Figure 5. The flow chart of system software.

5. Test result

FSQ26 signal analyzer (RS Company) was adopted for testing. The measured spectrum of the local oscillator is shown in figure 6. Figures 6(a), 6(b) and 6(c) show the spectrum of frequency 5600 MHz, 5608 MHz, and 5800 MHz with the "span" of analyzer set to 200 MHz, 5 MHz, and 5 GHz. It can be seen that synthetic frequency range covers from 5600 MHz to 5800 MHZ, with 8 MHz step. Figure 6(d) shows the spectrum of the frequency 5700 MHz with the analyzer "span" set to 20 kHz, it can be seen that the phase noise is -103.33 dBc/ Hz/ 10 KHz. The local oscillator has been tested by high and low temperature experiments. According to the test results, the frequency stability of frequency 5700MHz is 2.3ppm. The measured size of the circuit board is 115×45 mm. It can be seen that all indicators meet the system requirements.

IOP Conf. Series: Earth and Environmental Science **114** (2018) 012012 doi:10.1088/1755-1315/114/1/012012

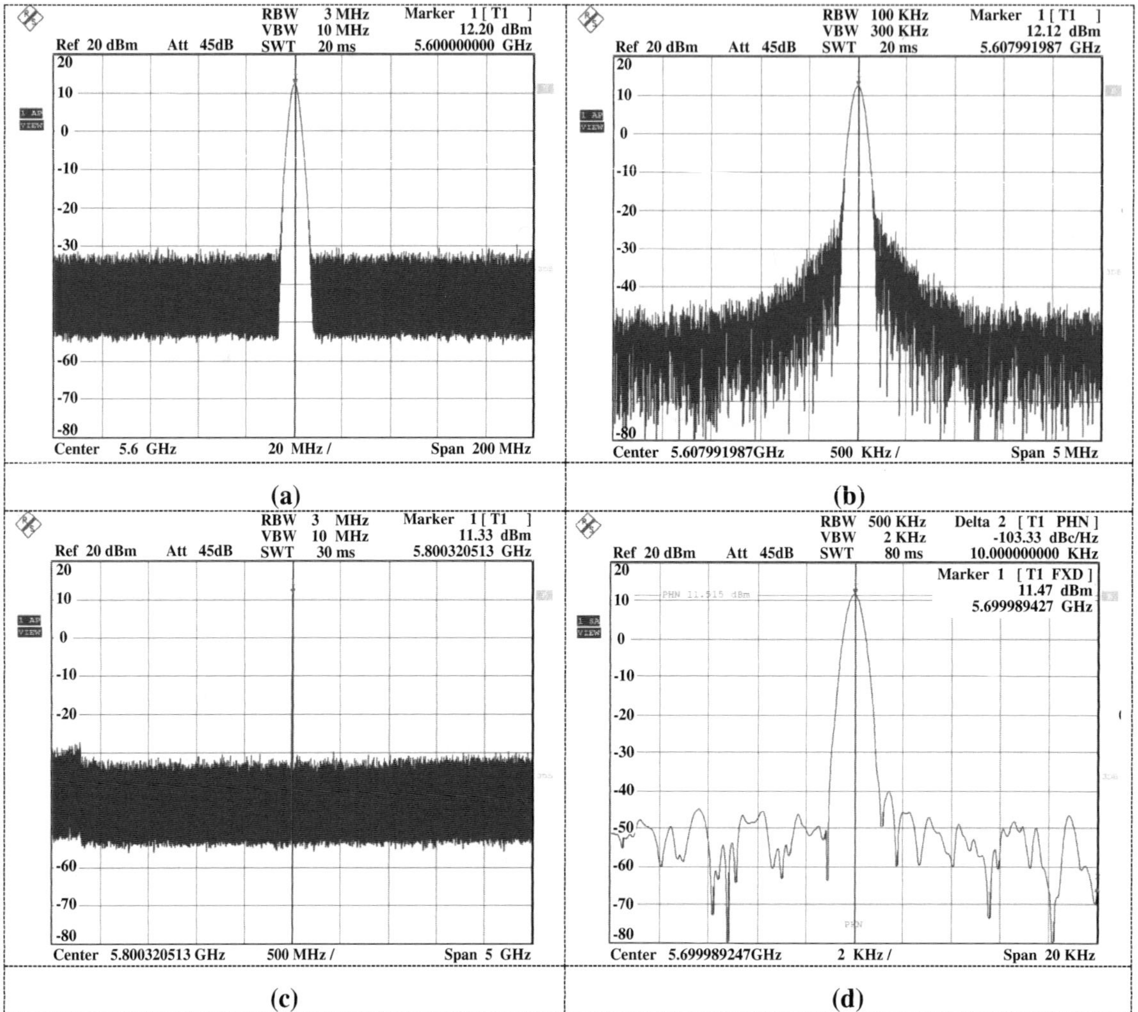

Figure 6. Test spectrum of the local oscillator by FSQ26 signal analyser of different frequency and span, (a) frequency=5600MHz and span=200MHz, (b) frequency=5608MHz and span=5MHz, (c) frequency=5800MHz and span=5GHz, (d) frequency=5700MHz and span=20KHz.

6. Conclusions

The local oscillator has been verified by flight tests. The results show that the design is reasonable, the indicators meet the system requirements, and the work is stable and reliable, which fully meet the requirements of the use of UAV data link telecontrol transmitter. This design can be applied to other related projects, with strong engineering practicability.

7. References

[1] Peizhang Wang 2016 Present Situation and Progress on Microwave Miniaturized Frequency Synthesizers Source *J. Military Communications Technology* **37** 35-39

[2] Saeed Golestan, Josep M Guerrero and Juan Vasquez 2017 A PLL-Based Controller For Grid-Connected Power Converters. IEEE Transactions on Power Electronics PP 1

[3] Sung-Geun Kim, Jinsoo Rhim and Dae-Hyun Kwon 2016 A Low-Voltage PLL With a Supply-Noise Compensated Feedforward Ring VC*O J.IEEE. TCAS II: Express Briefs* **63** 548-52

[4] ADI 2017 *HMC834LP6GE Datasheet* (Norwood: Analog Devices,inc)

[5] Yong Zhou, Xiaobo Li and Guoling Qin 2014 Design of Low Phase Noise and Low Stray Frequency Sources Based On HMC833 *Shipboard Electronic Countermeasure* **37** 114-7

[6] Yujian Pan, Xiaofa Zhang and Naichang Yuan 2012 Design of C-band broadband frequency source and its test system *Application of Electronic Technique* **38** 85-7

[7] Nagarajan Mahalingam, Yisheng Wang and Bharatha Kumar Thangarasu 2017 A 30-GHz Power-Efficient PLL Frequency Synthesizer for 60-GHz Applications *IEEE Transactions on Microwave Theory and Techniques* **PP** 1-11

[8] Kyoung-Jun Lee, Jong-Pil Lee and Dongsul Shin 2013 A novel grid synchronization PLL method based on adaptive low-pass notch filter for grid-connected PCS *IEEE Transactions on Industrial Electronics* **61** 292-301

[9] Kuba Raczkowski, Nereo Markulic and Benjamin Hershberg 2015 A 9.2–12.7 GHz wideband fractional-N subsampling PLL in 28 nm CMOS with 280 fs RMS jitter *IEEE Journal of Solid-State Circuits* **50** 1203-13

[10] Fran Gonzalez-Espin, Emilio Figueres and Gabriel Garcera 2012 An Adaptive Synchronous-Reference-Frame Phase-Locked Loop for Power Quality Improvement in a Polluted Utility Grid *IEEE Transactions on Industrial Electronics* **59** 2718-31

[11] Yujian Pan, Xiaofa Zhang and Naichang Yuan 2011 Design of low phase noise and low spurious frequency source based on HMC830 *Electronic Design Engineering* **19** 180-6

IOP Conf. Series: Earth and Environmental Science **114** (2018) 012013 doi:10.1088/1755-1315/114/1/012013

Techniques for Computation of Frequency Limited H_∞ Norm

Shafiq Haider, Abdul Ghafoor, Muhammad Imran and Malik Fahad Mumtaz

National University of Sciences and Technology (NUST), Islamabad, Pakistan
E-mail: m.imran@mcs.edu.pk

Abstract. Traditional H_∞ norm depicts peak system gain over infinite frequency range, but many applications like filter design, model order reduction and controller design etc. require computation of peak system gain over specific frequency interval rather than infinite range. In present work, new computationally efficient techniques for computation of H_∞ norm over frequency limited interval are proposed. Proposed techniques link norm computation with maximum singular value of the system in limited frequency interval. Numerical examples are incorporated to validate the proposed concept.

1. Introduction

Consider a continuous linear time invariant dynamical system:

$$\begin{aligned} \dot{x} &= Ax + Bu \\ y &= Cx + Du \end{aligned} \tag{1}$$

where $A \in R^{\{n\times n\}}, B \in R^{\{n\times m\}}, C \in R^{\{p\times n\}}, D \in R^{\{p\times n\}}$ with n= system order, m= number of inputs and p= number of outputs. The system (1) is called stable if all eigenvalues of A have negative real parts. The transfer matrix of (1) given as $G(j\omega) = C(j\omega\, I - A)^{\{-1\}}B + D$ is of order $p \times m$. For system (1), among many variants of norms, H_∞ norm is the most popular and extensively used for system analysis and design applications like model order reduction, cost function minimization, filter and controller design.

Definition 1: For stable system (1), the H_∞ norm of the system over infinite frequency range is given by [1]:

$$\begin{aligned} ||G(j\omega)||_{\{H_\infty\}} &= sup_{\{\omega\in R\}}\, \sigma_{\{max\}}(G(j\omega)) \\ ||G(j\omega)||_{\{H_\infty\}} &= sup_{\{\omega=[-\infty,+\infty]\}}\sigma_{\{max\}}(G(j\omega)) \end{aligned} \tag{2}$$

where $\sigma_{\{max\}}(.)$ is the largest singular value (SV) of the system over infinite frequency interval.

Using Definition 1, infinity norm can be computed via largest SV of the system transfer matrix. Many techniques that link norm with SV computation have been developed such as SV computation for time delay system [2], Routh table [3], characterization of polynomial [4], state space formulation [5], bisection [1], [6] etc. In [1] and [6] imaginary eigenvalues of Hamiltonian matrix are computed and linked with largest SV the system. Also in [7] extremum SVs are computed for strong $H_{\{\infty\}}$ norm computation. Similarly, largest SV can be computed using two sided Jacobi's method and Householder bidiagonalization method. Jacobi's method apply plane rotations to diagonalize transfer matrix [8]. Largest diagonal entry is the largest SV of the system. Although two sided Jacobi's method is computationally loaded, it is optimally accurate and guarantee stability even if transfer matrix

elements contain small relative errors [9]. Householder method (computationally less complex but less accurate as compared to Jacobi's technique) biadiagonalize transfer matrix by applying Householder transformation. Further it diagonalize system with orthogonal projections to yield largest SV [9].
Existing schemes [1]-[7] by definition, compute norm over infinite frequency range. However, many applications like frequency limited model reduction [10]-[13], filter design [14], signal reconstruction [15] etc. require analysis or design for the system in limited frequency interval. To the author's knowledge, no scheme to compute frequency limited H_∞ norm appear in literature (although concept of frequency limited $H_{\{2\}}$ norm has been developed [16]). Therefore in present work, techniques to compute frequency limited infinity norm via largest SV (by introducing frequency limited versions of two sided Jacobi and Householder methods for system (1)) are proposed.

2. Proposed techniques

Definition 2: Given the stable system (1), frequency limited infinity norm is defined as:

$$||G(j\omega)||_{\{H_{\{\infty\}},\delta\}} = sup_{\{\omega\in\delta\}}\,\sigma_{\{max,\delta\}}(G(j\omega)) \tag{3}$$

where $\delta = [-\omega_i, -\omega_{\{i-1\}}]U[\omega_{\{i-1\}}, \omega_i]$, $\omega_{\{i-1\}}$ is the lower frequency and ω_i is the higher frequency, U represents union and i is the frequency interval index.
Remark 1: When $\delta = [-\infty, +\infty]$, $\sigma_{\{max,\delta\}}(G(j\omega)) = \sigma_{\{max\}}(G(j\omega))$.
Remark 2: δ may contain multiple frequency intervals as $\delta = [-\omega_{\{i\}}, -\omega_{\{i-1\}}]U\ldots U[-\omega_2, -\omega_1]U[\omega_1, \omega_2]\ldots U[\omega_{\{i-1\}}, \omega_i] = \delta_1\, U\ldots U\delta_{\{i-1\}}$. Consequently, $||G(j\omega)||_{\{H_\infty,\delta\}} = max\,(\,||G(j\omega)|_{\{H_\infty,\delta_1\}}, ||G(j\omega)||_{\{H_\infty,\delta_2\}}, \ldots ||G(j\omega)||_{\{H_\infty,\delta_{\{i-1\}}\}})$.

2.1. *Two Sided Jacobi's Technique for Frequency Limited H-Infinity Norm Computation:*

Definition 3: Jacobi's two sided transformation is defined as [17]:

$$Z = J^T G(j\omega_\delta)J \tag{4}$$

where $G(j\omega_\delta) = C(j\omega_\delta\, I - A)^{\{-1\}}B + D$, $\omega_\delta \in \delta$, J is the Jacobi rotation matrix given by:

$$J = \begin{bmatrix} c & s \\ -s & c \end{bmatrix} \tag{5}$$

where $c = cos(\theta)$, $s = sin(\theta)$ and θ is the rotation applied. On applying Jacobi rotation to (p,q) block of $G(j\omega_\delta)$, we obtain:

$$\begin{bmatrix} Z_{\{pp\}} & Z_{\{pq\}} \\ -Z_{\{qp\}} & Z_{\{qq\}} \end{bmatrix} = \begin{bmatrix} c & s \\ -s & c \end{bmatrix}^T \begin{bmatrix} G(j\omega_\delta)_{\{pp\}} & G(j\omega_\delta)_{\{pq\}} \\ G(j\omega_\delta)_{\{qp\}} & G(j\omega_\delta)_{\{qq\}} \end{bmatrix} \begin{bmatrix} c & s \\ -s & c \end{bmatrix} \tag{6}$$

To reduce to diagonal form, set $Z_{\{pq\}} = Z_{\{qp\}} = 0$. Manipulation and comparison yield:

$$G(j\omega_\delta)_{\{pq\}}\,(c^2 - s^2\,) + (G(j\omega_\delta)_{\{pp\}} - G(j\omega_\delta)_{\{qq\}})cs = 0$$
$$\frac{G(j\omega_\delta)_{\{qq\}} - G(j\omega_\delta)_{\{pp\}}}{G(j\omega_\delta)_{\{pq\}}} = \frac{(c^2 - s^2)}{2cs} = cot(2\theta) = \zeta$$

Let $t = tan\,(\theta)$ satisfy $t^2 + 2\zeta\, t - 1 = 0$ whose solution give:

$$t = -sig(\zeta)(|\zeta| + \sqrt{(\{1+\zeta^2\,\})} = sig(\zeta)/(|\zeta| + \sqrt{(\{1+\zeta^2\,\})})$$
$$c = 1/\sqrt{(\{1+\zeta^2\,\})}s = ct \tag{7}$$

where sig is the signum function, c and s are invoked in (5) and (4) to obtain diagonal form and consequent largest SV. In following summary of method discussed above is presented.

Algorithm 1: Frequency Limited H_∞ Norm Computation using Jacobi's Technique

Input: Given the system (1), frequency interval $\delta = [\omega_1, \omega_2]$ and tolerance $\in$:

Output: Frequency limited H_∞ Norm.

for δ

$|G(j\backslash omega\_\backslash delta)| = |C(j\backslash omega\_\backslash delta\, I - A)\text{^}\{-1\}B + D|$

repeat

for all pairs $p < q$

Compute J from (7) and (5)

Compute Z from (4)

Update $|G(j\omega_\delta)|$ *by solving* (6)

Compute $\sigma_{\{max,\delta\}}(G(j\omega_\delta)) = max(G(j\omega_\delta)_{\{pp\}}, G(j\omega_\delta)_{\{qq\}}$

until(all $\dfrac{|G(j\omega_\delta)_{\{pq\}}|}{(G(j\omega_\delta)_{\{pp\}} G(j\omega_\delta)_{\{qq\}})^{\{\frac{1}{2}\}}} \leq \in)$

end for

2.2. *Householder Transformation*

Definition 4: Householder transformation for bidiagonalization is defined as [18]:

$$B(j\omega_\delta) = H_k G(j\omega_\delta) O_k \tag{8}$$

where $G(j\omega_\delta) = C(j\omega_\delta I - A)^{\{-1\}}B + D$ is of order $p \times m, \omega_\delta \in \delta, k$ is the column or row of $G(j\omega_\delta)$ whose selected elements are to be zeroed out, H_k is premultiplier Householder that successively zero out elements below (k,k) entries of $G(j\omega_\delta)$ computed by:

$$H_k = I - 2w_{\{H_k\}} w_{\{H_k\}}^T \tag{9}$$

where

$$\begin{aligned} w_{\{H_k\}} &= v_{\{H_k\}} / \left\| v_{\{H_k\}} \right\|_2 \\ v_{\{H_k\}} &= (0, \ldots, 0, a_{\{kk\}}^k - \alpha_{\{H_k\}}, a_{\{k+1,k\}}^k, \ldots, a_{\{pk\}}^k) \\ \alpha_{\{H_k\}} &= -sig(a_{\{kk\}}^k) || (0, \ldots, 0, a_{\{kk\}}^k, a_{\{k+1,k\}}^k, \ldots, a_{\{pk\}}^k) || \end{aligned}$$

$a_{\{kk\}}$ is the (k,k) element of $G(j\omega_\delta)$ and O_k is post multiplier that successively zero out transfer matrix elements past $(k, k+1)$ entries to yield bidiagonal form of system matrix.

$$O\_k = I - 2w\_\{O\_k\}\text{^}Tw\_\{O\_k\} \tag{10}$$

where

$$\begin{aligned} w_{\{O_k\}} &= v_{\{O_k\}} / \left\| v_{\{O_k\}} \right\|_2 \\ v_{\{O_k\}} &= (0, \ldots, 0, a_{\{kk+1\}}^k - \alpha_{\{O_k\}}, a_{\{k,k+2\}}^k, \ldots, a_{\{km\}}^k) \\ \alpha_{\{O_k\}} &= -sig(a_{\{kk+1\}}^k) || (0, \ldots, 0, a_{\{kk+1\}}^k, a_{\{k,k+2\}}^k, \ldots, a_{\{km\}}^k) || \end{aligned}$$

Remark 3: Householder transformation maps one vector subspace into another by preserving the norms of initial and resulting vectors. Moreover $H_k = H_k^T, H_k^{\{-1\}} = H_k^T, H_k^2 = I$.

Remark 4: Householder bidiagonalization is a successive process i.e. $H_1 G(j\omega_\delta) = S_1, S_1 O_1 = S_2, H_2 S_2 = S_3, \ldots, S_m O_m = B(j\omega_\delta)$ and

$$G(j\omega\_\delta) = (H_1...H_p)B(j\omega_\delta)(O_m...O_1).$$

After bidiagonal form $B(j\omega_\delta)$ is obtained, it's orthogonal matrices P and Q are computed such that $\Sigma = P^T B(j\omega_\delta)Q$ is diagonal and nonnegative. Largest diagonal entry qualify as infinity norm in given frequency interval. The columns of P and Q are right and left singular vectors respectively.

Algorithm 2: Frequency Limited H_∞ Norm Computation using Householder Technique
Input: Given the system (1), frequency interval $\delta = [\omega_1, \omega_2]$:
Output: Frequency limited H_∞ Norm.

for δ
  $G(j\omega_\delta) = C(j\omega_\delta I - A)^{\{-1\}}B + D$
  Set $S_{\{k-1\}} = S_0 = G(j\omega_\delta)$
  for $k = 1{:}m$
    Compute H_k from (9)
    Compute $S_{\{kH\}} = H_k S_{\{k-1\}}$
    Compute O_k from (10)
    Apply post multiplier $S_k = S_{\{kH\}}O_k$
  end for
  Set $B(j\omega_\delta) = S_k$
  Diagonalize to obtain $\Sigma = P^T B(j\omega_\delta)Q$
  Compute norm by $\sigma_{\{max,\delta\}} = max(\Sigma)$
end for

Remark 5: Techniques A and B can be used for computation of H_∞ norm for discrete time systems as well.

Remark 6: The proposed techniques A and B are applicable to arbitrarily constructed matrices. However as most applications of frequency limited H_∞ norm are related to systems having certain physical interpretation, therefore posteriori constructed system matrices are considered in present work.

3. Illustrative examples

Proposed techniques are applied to many continuous and discrete time systems out of which results for few examples are presented.

3.1. Continuous Time Systems

Example 1: Consider following $6^{\{th\}}$ order SISO system [19]:

$$A = \begin{bmatrix} 0 & 0 & 0 & 1 & 0 & 0 \\ 0 & 0 & 0 & 0 & 1 & 0 \\ 0 & 0 & 0 & 0 & 0 & 1 \\ -5.4545 & 4.5455 & 0 & -.0545 & .0455 & 0 \\ 10 & -21 & 11 & .1000 & -.2100 & .1100 \\ 0 & 5.5000 & -6.5000 & 0 & .0550 & -.0650 \end{bmatrix}$$

$$B = \begin{bmatrix} 0 & 0 & 0 & .0909 & .4 & -.5 \end{bmatrix}^T$$

$$C = \begin{bmatrix} 2 & -2 & 3 & 0 & 0 & 0 \end{bmatrix}, D = 0$$

Example 2: Consider another $6^{\{th\}}$ order MIMO system [20]:

$$A = \begin{bmatrix} -20.02 & -0.124 & -0.203 & -0.254 & 0.203 & 0.3057 \\ 3.967 & -0.165 & 1.017 & 1.272 & -1.017 & -1.526 \\ -0.279 & -1.399 & -7.118 & -2.647 & 0.117 & 0.1766 \\ -0.349 & -1.749 & 0.9872 & -3.766 & 3.013 & 4.519 \\ 0.2798 & 1.399 & 0.2253 & 0.2816 & -5.225 & -3.338 \\ 0.4196 & 2.098 & 3.134 & 3.917 & -1.134 & -4.7 \end{bmatrix}$$

$$B = \begin{bmatrix} 2 & 1.67e-016 \\ 2.665e-015 & 8.352e-016 \\ 0.8296 & 2 \\ 1.037 & 1.665e-016 \\ -0.8296 & 2 \\ -1.244 & -2.22e-016 \end{bmatrix}$$

$$C = \begin{bmatrix} 0.2378 & 1.189 & 0.6226 & 0.6533 & -0.122 & -0.183 \\ 0.3584 & 0.5419 & 0.5319 & 0.6648 & -0.031 & 0.7022 \end{bmatrix}$$

$$D = \begin{bmatrix} 0 & 0 \\ 0 & 0 \end{bmatrix}$$

3.2. Discrete Time Systems

Example 3: Consider following $4^{\{th\}}$ order discrete time SISO system [21]:

$$A_d = \begin{bmatrix} -1.1000 & 0.0100 & 0.2750 & 0.0600 \\ 1.0000 & 0 & 0 & 0 \\ 0 & 1.0000 & 0 & 0 \\ 0 & 0 & 1.0000 & 0 \end{bmatrix}$$

$$B_d = \begin{bmatrix} 1 & 0 & 0 & 0 \end{bmatrix}^T$$

$$C_d = \begin{bmatrix} 1 & 0 & 0 & 0 \end{bmatrix}, D_d = 0$$

Example 4: Consider following discrete time $6^{\{th\}}$ order system [21]:

$$A_d = \begin{bmatrix} -0.8750 & -0.75 & -0.5 & -0.3 & -0.25 & -0.1 \\ 1.0000 & 0 & 0 & 0 & 0 & 0 \\ 0 & 1 & 0 & 0 & 0 & 0 \\ 0 & 0 & 1 & 0 & 0 & 0 \\ 0 & 0 & 0 & 1 & 0 & 0 \\ 0 & 0 & 0 & 0 & 1 & 0 \end{bmatrix}$$

$$B_d = \begin{bmatrix} 1 & 0 & 0 & 0 & 0 & 0 \end{bmatrix}^T$$

$$C_d = \begin{bmatrix} 0.25 & 1.25 & 1.75 & 2 & 2.5 & 0.25 \end{bmatrix}, D_d = 0$$

SV plot for systems are shown in Fig. 1, Fig. 2, Fig. 3 and Fig. 2 respectively and computed infinity norms are given in Table 1 and Table 2 respectively, for various frequency intervals. The computed

IOP Conf. Series: Earth and Environmental Science **114** (2018) 012013 doi:10.1088/1755-1315/114/1/012013

norms match with the maximum SV in given frequency interval that certify the correct development of the proposed techniques.

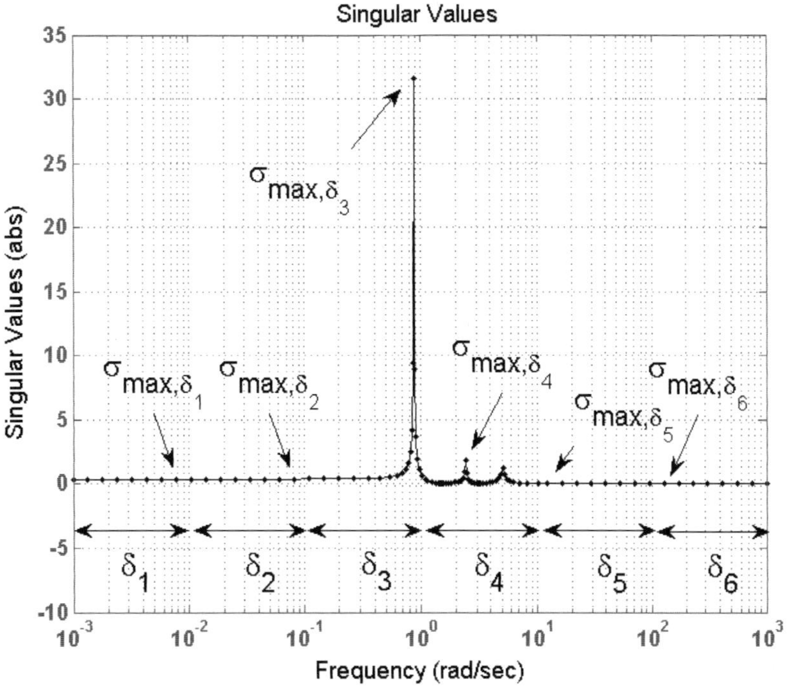

Figure 1. SV plot for example 1.

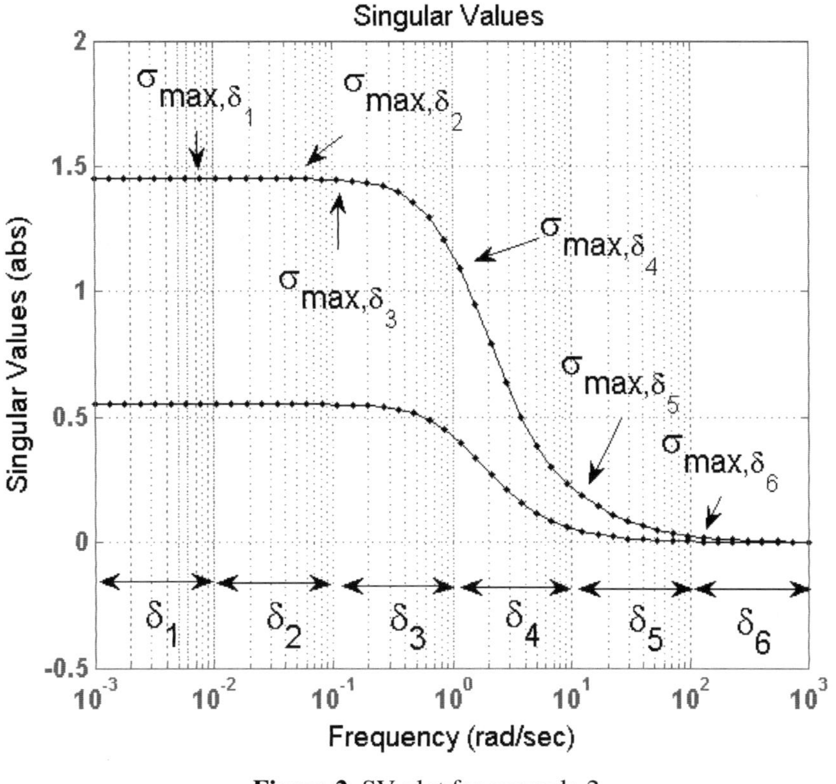

Figure 2. SV plot for example 2.

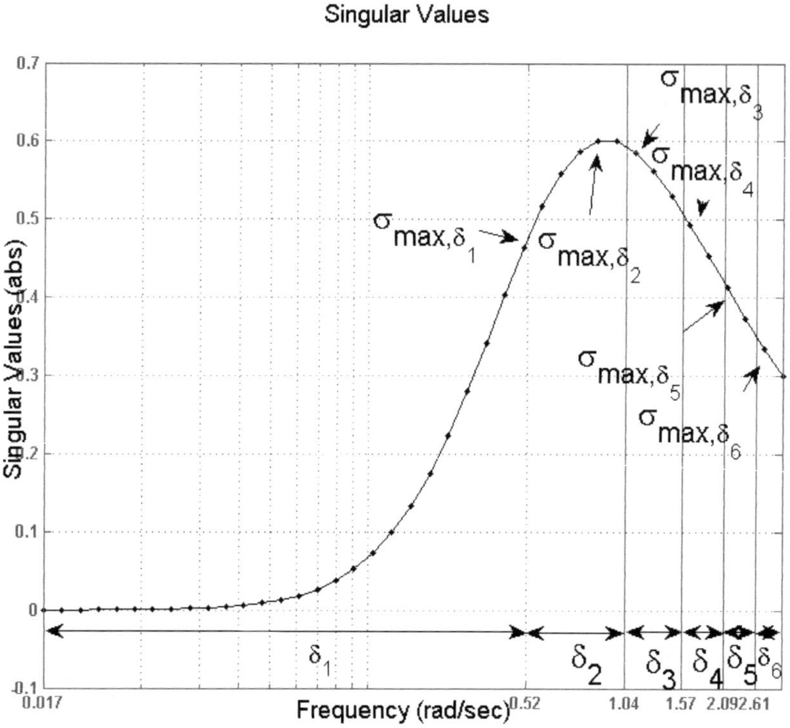

Figure 3. SV plot for example 3.

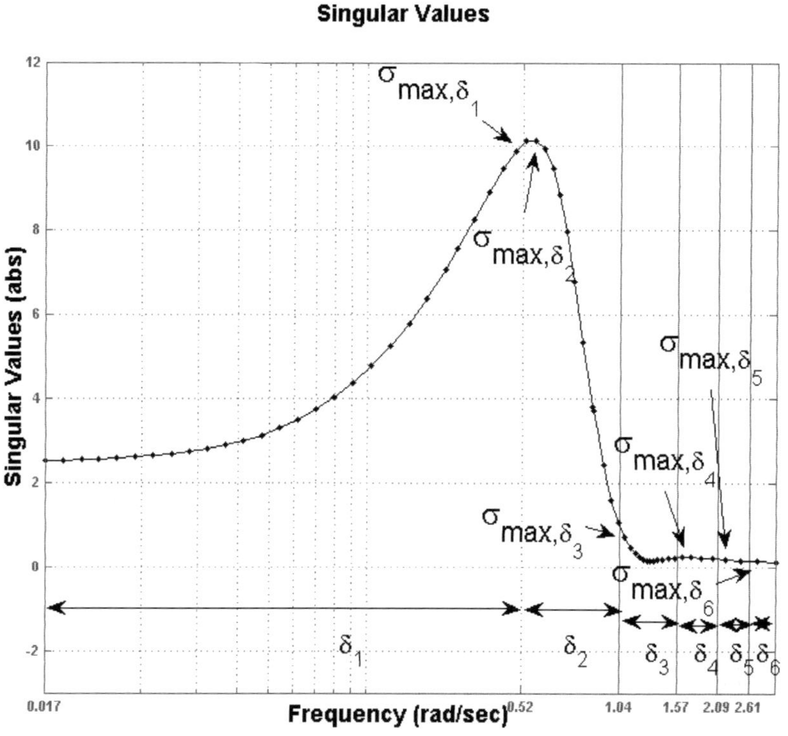

Figure 4. SV plot for example 4.

IOP Conf. Series: Earth and Environmental Science **114** (2018) 012013 doi:10.1088/1755-1315/114/1/012013

Table 1. Frequency limited infinity norms for continuous time systems.

Frequency Interval	Frequency Limited Norm	
	Example 1	Example 2
$\delta_1 = [10^{-3}, 10^{-2}]$	0.3810	1.4512
$\delta_2 = [10^{-2}, 10^{-1}]$	0.3847	1.4512
$\delta_3 = [10^{-1}, 10^{0}]$	31.5564	1.4465
$\delta_4 = [10^{0}, 10^{1}]$	1.8019	1.1542
$\delta_5 = [10^{1}, 10^{2}]$	0.0272	0.2166
$\delta_6 = [10^{2}, 10^{3}]$	2.1227e-04	0.0261
$Overall = [10^{-3}, 10^{3}]$	31.5564	1.4512

Table 2. Frequency limited infinity norms for discrete time systems.

Frequency Interval	Frequency Limited Norm	
	Example 3	Example 4
$\delta_1 = [\pi/180, \pi/6]$	0.4747	10.1704
$\delta_2 = [\pi/6, \pi/3]$	0.6020	10.1691
$\delta_3 = [\pi/3, \pi/2]$	0.5941	1.0026
$\delta_4 = [\pi/2, 2\pi/3]$	0.5060	0.2333
$\delta_5 = [2\pi/3, 5\pi/6]$	0.4177	0.2000
$\delta_6 = [5\pi/6, \pi]$	0.3502	0.1495
$Overall = [\pi/180, \pi]$	0.6020	10.1704

4. Concluding remarks

In order to emphasize system analysis, design and optimization in limited frequency interval, two techniques to compute frequency limited infinity norm are proposed. In both techniques namely Jacobi's and Householder, largest SV of the system is computed at each frequency point and maximum of these values is taken over limited interval. The computed values match with maximum SV in limited frequency interval that certify the successful development of the proposed schemes.

5. References

[1] S. Boyd, V. Balakrishnan, and P. Kabamba, A bisection method for computing the H_∞ norm of a transfer matrix and related problems, *Math. Cont. Sig. Sys., New York*, vol. 2, no. 3, pp. 207–219, 1989.
[2] S. Gumussoy, and W. Michiels, Computing H¥ norm of time delay systems, *Deci. Cont. Conf.*, pp. 263–268, 2009.
[3] A. G. Aghdan, A method to obtain the infinity-norm of systems using the Routh table, *Proc.* IEEE *Aut. Cong., Bud., Hungary*, pp. 1–6, 2006.
[4] M. N. Belur, and C. Paragman, Efficient computation of the H_∞ norm, *Proc. IFAC. Symp. Rob. Cont. Des., Toulsouse*, 2006.
[5] J. Kuster, and E. George, H_∞ norm calculation via a state space formulation, *VirginiaTech*, 2013.
[6] S. Boyd, V. Balakrishnan, and P. Kabamba, On computation of the H_∞ norm of a transfer matrix, *Proc. Amer. Cont. Conf., Atlanta, Georgia*, pp. 396–397, 1988.
[7] G. Suat, and M. Wim, Computation of extremum SV's and the strong H_∞ norm of SISO time delay systems, *Automatica*, vol. 54, pp. 266-271, 2015.
[8] L. Jim, Lecture notes on Jacobi Methods, *Standford Uni., California*, 2010-11.

[9] D. James, and V. Kresimir, Jacobi's Method is more accurate than QR, *Comp. Sci. Dept. Tech.* Rep., *Courant Inst., New York*, 1989.
[10] S. Sahlan, A. Ghafoor, and V. Sreeram, A new method for the model reduction technique via a limited frequency interval impulse response Gramian, *Mathe. Comp. Mod.*, vol. 55, no. (34), pp. 10341040, 2012.
[11] D. Wang, and A. Zilouchian, Model reduction of discrete linear system via frequency domain balanced realization, *IEEE Trans. Circ. Sys. I*, vol. 47, no. 6, pp. 830-837, 2000.
[12] M. Imran, A. Ghafoor, S. Akram, and V. Sreeram Limited frequency interval gramians based model reduction for nonsingular generalized systems, *Aust. Cont. Conf.,*, pp. 441-444, 2013.
[13] X. Du, V. Sreeram, V. Togneri, A. Ghafoor, and S. Sahlan, A frequency limited model reduction technique for linear discrete systems, *Aust. Cont. Conf., Perth, Australia*, pp. 421–426, 2013.
[14] G. Meinsma, and H. S. Shekhawat, Frequency-truncated system norms, *Automatica*, vol. 47, no. 8, pp. 1842-1845, 2011.
[15] G. Meinsma, and H. S. Shekhawat, Truncated norms and limitations on signal reconstruction, Int. Symp. Mathe. the. Netw. Sys., pp. 1135-1140, 2010.
[16] P. Vuillemin, C. Paussot-vassel, and D. Alazand, A spectral expression for the frequency limited H2 norm of LTI dynamical systems with higher order poles, *Proc. Euro. Cont. Conf.,* pp. 55-60, 2014.
[17] P. B. Richard, T. L. Franklin, and V. L. Charles, Computation of SV decomposition using mesh-connected processors, *Jour. VLSI Comp. Sys.*, vol. 1, no. 3, pp. 242-270, 1982.
[18] D. John, Derivations for linear algebra and optimization, *Comp. Sci. Div., Berkeley,* 2007.
[19] W. Gawronski, and J. N. Juang, Model reduction in limited time and frequency intervals, *Int. Jour. Sys. Sci.,* vol. 21, no. 2, pp. 349–376, 1990.
[20] D. K. Kranthi, S. K. Nagar, and S. K. Bharadwaj, Model order reduction of SISO and MIMO systems based on genetic algorithm, *Int. Conf. Aut. Robo. Cont. Sys*., pp. 1-8, 2010.
[21] M. Imran, and A. Ghafoor, Stability Preserving Model Reduction Technique and Error Bounds Using Frequency-Limited Gramians for Discrete-Time Systems, *IEEE Trans. Circ. Sys. II,* vol. 61, no. 9, pp. 716–720, 2014.

IOP Conf. Series: Earth and Environmental Science **114** (2018) 012014 doi:10.1088/1755-1315/114/1/012014

Research on Abnormal Detection Based on Improved Combination of K - means and SVDD

Hao Xiaohong[1], **Zhang Xiaofeng**[2]

[1]School of Electrical Engineering and Information Engineering, Lanzhou University of Technology, Lanzhou 730050, China
[2]School of Computer and Communication, Lanzhou University of Technology, Lanzhou 730050, China

Abstract. In order to improve the efficiency of network intrusion detection and reduce the false alarm rate, this paper proposes an anomaly detection algorithm based on improved K-means and SVDD. The algorithm first uses the improved K-means algorithm to cluster the training samples of each class, so that each class is independent and compact in class; Then, according to the training samples, the SVDD algorithm is used to construct the minimum superspheres. The subordinate relationship of the samples is determined by calculating the distance of the minimum superspheres constructed by SVDD. If the test sample is less than the center of the hypersphere, the test sample belongs to this class, otherwise it does not belong to this class, after several comparisons, the final test of the effective detection of the test sample.In this paper, we use KDD CUP99 data set to simulate the proposed anomaly detection algorithm. The results show that the algorithm has high detection rate and low false alarm rate, which is an effective network security protection method.

1. Introduction

With the rapid development of computer technology and Internet technology, intrusion has become a major threat to network security. In order to deal with network intrusion, network security experts in the field of a number of technical means, such as: firewall, data encryption and secure routing [1]. However, relying solely on a single static security technology such as firewalls can not resist complex and complex network attacks. As a result, experts in the field of network security proposed a network intrusion detection technology.

Intrusion detection technology is divided into two categories: misuse detection and anomaly detection [2]. Misuse detection refers to the use of specific types of intrusion behavior to create a feature library, when detected with the characteristics of the matching behavior is considered intrusion behavior [3]. But the method has a high false negative rate. Exception detection refers to the use of normal user behavior learning modeling, get the normal situation of the probability model [4]. Detection, the user behavior and the characteristics of the library to compare, the greater the deviation is regarded as intrusion. Based on the statistical anomaly detection is the traditional intrusion detection technology, it can only small-scale network detection, the face of large-scale network of its detection efficiency is low. Data mining technology can analyze the Internet data, from which to tap the useful information, and use this information to detect abnormal intrusion and known intrusion has become the most important network intrusion detection tool [5].

Wenke Lee et al [6] applied the data mining technology to the intrusion detection system for the first time. Through the use of data mining method to analyze a large number of audit data, through this method can get more accurate results, and its learning efficiency will be improved, can effectively improve Intrusion detection system accuracy and expand its scope of application, but the use of this method requires a larger storage space. In order to solve this problem, Keerthi et al [7] proposed a decomposition algorithm that transforms the quadratic programming problem into a series of small-scale quadratic programming problems, which improves the learning rate and reduces the memory space requirements without affecting the classification accuracy, and A SMO algorithm is proposed, which can solve the optimal solution of each sub-problem and accelerate the convergence speed of the algorithm. However, this method increases the computational complexity. In order to further improve the learning speed of the algorithm, Collobert et al [8]. Proposed a parallel learning algorithm for sample set segmentation, which divides the sample set into P sub-samples by using the principle of divide and conquer, so as to improve the detection efficiency of the algorithm. However, since the P subsets are randomly divided, the accuracy of the classification is reduced, and the addition of additional rules increases the complexity of the algorithm. However, since the P subsets are randomly divided, the accuracy of the classification is reduced, and the addition of additional rules increases the complexity of the algorithm. In order to reduce the computational complexity, Cataltepe Z et al [9]. Proposed a semi-supervised decision tree algorithm for online feature selection, which uses online clustering to summarize the available network data, using extended cluster features to represent clusters, The characteristics of the original feature, but also the characteristics of the relationship between the description of the cluster, each cluster is marked as abnormal or normal by the user, and then based on the information to train the decision tree, according to the output of the decision tree to mark the input of new data. Although the computational complexity of the algorithm is reduced, the detection rate is still not high. In this paper, an improved anomaly detection algorithm combining K-means and SVDD is proposed. The basic idea of this algorithm is to cluster each category by K-means clustering algorithm first, and then build multiple hyperspheres from SVDD according to the result of clustering. By calculating the sample data to the minimum superspheres constructed by SVDD Distance to determine the affiliation of the sample. Finally, this paper uses the KDD CUP99 data set to simulate the proposed detection algorithm.

2. K-means algorithm and its improvement

2.1 K-means algorithm

K-means clustering algorithm is a typical partitioning method. It divides n objects into K clusters with K as input parameters, so that the clusters are independent and similar, and the similarity of clusters is The average of the objects in the cluster is calculated [10]. Although the computational complexity of K-means algorithm is low and easy to implement, there is a major drawback of this algorithm, which is sensitive to outliers [11]. Because a large extreme value of the object may significantly distort the data distribution, coupled with the K-means algorithm is to use the square error function to determine the convergence, making its sensitivity to the outlier increase, resulting in clustering results Inaccurate.

2.2 Improved K-means algorithm

In order to reduce the sensitivity of traditional K-means algorithm to outliers, an improved K-means algorithm is adopted in this paper. The basic idea of this algorithm is to represent the cluster by selecting the actual object in the cluster, and the remaining objects are allocated to the cluster with the most similar representative objects, so that the partitioning method can still be used to minimize all objects Corresponding to the degree of dissimilarity between the reference points [12]. At this point, the criterion function uses the absolute error criterion, which is defined as follows:

$$E = \sum_{i=1}^{k} \sum_{p \in c_i} \left| p - o_j \right|^2$$

Where E is the square error of all the objects in the data set, and the point in the p space, representing a given object in the cluster C_i, and o_j is the representative object in the cluster C_i.

The improved K-means algorithm has been iterated several times until each representative object becomes the actual center of its cluster. In the iterative process, as long as the clustering results continue to change, this iterative process will continue to use non-representative objects to replace the representative object. In order to determine whether the non-representative object o_{random} is the best representative of the current representative, the following four cases need to be considered for each non-object representative:

1) p currently belongs to the representative object o_j, and redistribute p to o_i if o_j is replaced by o_{random} as the new delegate object and p is closest to the other representing object $o_i(i \neq j)$.
2) p currently belongs to the representative object o_j, if the o_j is replaced by o_{random} as a new representative object, and p distance o_{random} is nearest, then p is reassigned to o_{random}.
3) p currently belongs to the representative object o_i, $i \neq j$. If o_j is replaced by o_{random} as a new representative object, and p is still closest to o_i, the membership of the object does not change.
4) p currently belongs to the representative object o_i, $i \neq j$. If o_j is replaced by o_{random} as a new representative object, and p is closest to o_{random}, then p is reassigned to o_{random}.

The difference in absolute error has an effect on the cost function when each reallocation occurs. Thus, if the current representative object is replaced by a non-representative object, the cost function calculates the difference in absolute error. The total cost of the exchange is the sum of the costs of all non-representative objects. When the total cost is negative, the actual absolute error E will be reduced, o_j can be replaced by o_{random}; when the total cost is positive, then the current representative of the object o_j is reasonable, in this iterative process will not change.

The improved K-means algorithm steps are as follows:

1) arbitrarily select K objects from D as the initial representative;
2) Repeat
3) assign each remaining object to the cluster represented by the nearest delegate object;
4) arbitrarily select a non-representative object o_{random};
5) calculate the total cost of representing the object o_j with the o_{random} switch;
6) If $S < 0$, then o_j is replaced by o_{random} to form a new set of K representative objects;
7) until the cluster no longer changes.

3. An Algorithm for Abnormal Detection Based on Improved K - means and SVDD

The anomalous detection algorithm based on improved K-means and SVDD proposed in this paper is based on two considerations [13]: First, in most cases, the normal data set is far redundant with abnormal data sets; Second, the imbalance between normal and abnormal data makes it possible to train the model when the training results may lead to deviation, affecting the test results. In this paper, we first use the improved K-means algorithm to cluster the training samples of each class so that each cluster is similar to each other, and then the result of clustering is used to train the SVDD algorithm. Clustering support vector, and finally all the individual support vector combination retraining to get the global optimal solution. Support vector domain description (SVDD) algorithm is a kind of support vector machine (One-Class SVM) proposed by TAX [14] and so on in the traditional support vector machine. The idea of the algorithm is to use the sample to be classified as a whole, to establish a closed area, so that all the samples belonging to the class all or as much as possible in this closed area, and do not belong to the class of exclusion In this closed area, the sample on the region is the support vector obtained by SVDD.

When a normal data set $\{x_1, x_2, \ldots x_n\}$ is given, which contains n data objects, as a training sample for constructing the One-class classifier, an attempt is made to find a minimum volume of the hypersphere (center a, radius R) so that all or as much as possible with x_i is contained within the supersphere [15]. In general, the training samples are more or less containing noise, so the above optimization results are more sensitive to these noise, lack of robustness [16]. In order to improve the robustness of the algorithm, the relaxation variable $\xi_i \geq 0$ is introduced for each sample x_i. Minimizing the volume of the hypersphere is the quadratic programming problem, that is minimizing $\min\left[R^2 + C\sum_i \xi_i\right]$ under the constraint condition $(x_i - a)^2 \le R^2 + \xi_i$.

Where C is a specified constant, its role is to control the degree of punishment on the wrong sample, to achieve the wrong proportion of the sample and the complexity of the algorithm between the compromise. Using the Lagrangian function to solve the minimum problem under the above constraints, we can get the dual problem of the original problem:

$$L = \sum_i a_i K(x_i, x_j) - \sum_i a_i a_j K(x_i, x_j)$$

In the above formula, the a_i should conform to the constraints: $\sum_i a_i = 1$ and $0 \le a_i \le C$, and $K(x,x)$ is the inner product of the above formula, so $K(x,x)$ should satisfy the kernel function of Mercer condition [17].

By the KKT condition, the sample corresponding to the $a_i = 0$ is in the hypersphere; the corresponding $0 \le a_i \le C$ sample is called the support vector on the hypersphere, which is regarded as an anomaly point with the partial sample corresponding to the $a_i > C$.

When a new sample Z is given, it is judged whether the sample belongs to the target sample, and the corresponding discriminant function is:

$$\|z-a\|^2 = K(z,z) - 2\sum_i a_i K(x_i, z) + \sum_{i,j} a_i a_j K(x_i, x_j) \le R^2$$

If the above discriminant function holds, then Z belongs to the target sample and belongs to that class; otherwise Z is not the target sample and does not belong to that class.

4. Simulation experiment

4.1 Experimental data set

In this paper, we use KDD CUP99 [18] data set to detect the proposed intrusion detection method. The data set includes four types of attack types: DoS, Probe, R2L, and U2R. In the KDD CUP99 data set, each network connection consists of 41 feature attributes and 1 flag attribute, of which there are 7 discrete feature attributes and 34 consecutive feature attributes [19]. In order to make the test results are universal, this paper uses three sets of data sets to carry out experiments, experimental data shown in Table 1.

Table 1 Experimental data

Data set	Normal	DoS	Probe	R2L	U2R
1	500	400	350	200	40
2	450	300	300	150	30
3	400	200	280	100	25

4.2 Experimental results and analysis

In this paper, the proposed algorithm is tested by simulation experiment. The experimental results are shown in Table 2.

Table 2 Comparison of experimental results

Atta	K-means	SVDD	Proposed

ck					algorithm	
type	Detect ion rate (%)	Fals e alar m rate (%)	Dete ctio n rate (%)	False alar m rate (%)	Detec tion rate (%)	Fals e alar m rate (%)
DoS	91.25	6.37	93.9 7	7.24	98.64	4.52
Pro be	93.74	5.78	96.2 1	5.78	97.81	3.85
R2L	73.51	25.2 4	84.6 2	24.3 6	89.42	20.2 1
U2 R	90.38	8.94	94.3 5	8.10	95.54	6.97

It can be seen from Table 2 that the improved K-means and SVDD proposed intrusion detection algorithm compared with the other two methods of DoS, Probe,U2R and other three types of attack has a high detection rate, but R2L this type of attack detection rate is low, false alarm rate of 20.21%. This is because the sample of this attack is less than other types of attacks, and many of the attacks in R2L have high similarity to normal data, so that some attacks are included in the normal type, resulting in a relatively low detection rate.

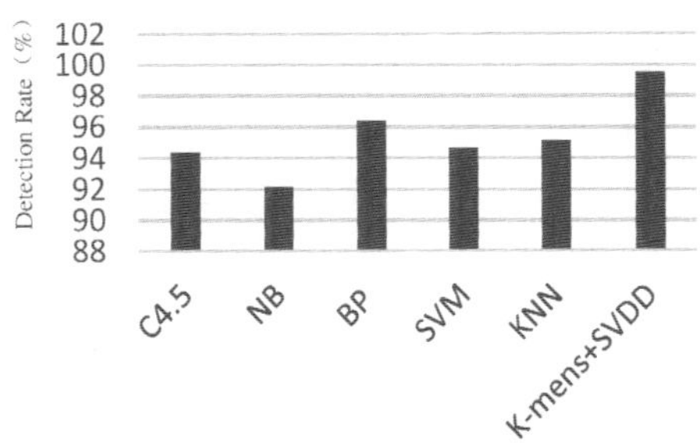

Figure 1. Comparison of detection rates for different algorithms

Figure 1 shows that the proposed intrusion detection algorithm combined with the improved K-means and SVDD is compared with the commonly used intrusion detection algorithm. It can be seen that the proposed method has obvious advantages over other detection methods in detecting detection rate. The algorithm based on BP neural network and the algorithm proposed in this paper are relatively close to the detection rate, but the former computational complexity is relatively high, so it takes a long training time.

5. Conclusion

In this paper, an improved K-means and SVDD algorithm is proposed to solve the problem of low detection rate and high false alarm rate in network intrusion detection. Firstly, the algorithm is clustered by the improved K-means clustering algorithm. Then, a number of superspheres are established by SVDD according to the result of clustering. By calculating the distance between the minimum superspheres constructed by SVDD Determine the affiliation of the sample, if the test sample within the hypersphere, the sample belongs to the class, otherwise it does not belong to the class.The experimental results show that the proposed algorithm has high detection efficiency and low false alarm rate for different attacks, which has obvious advantages over traditional intrusion detection methods. However, the intrusion detection algorithm proposed in this paper also has some shortcomings, such as when the attack type of sample data is less detection efficiency is not too ideal. Therefore, in the future research work, the use of a variety of classification algorithms to build intrusion detection system will be a hot research.

6. References

[1] Chetan R, Ashoka D V. Data mining based network intrusion detection system: A database centric approach[C]//Computer Communication and Informatics (ICCCI), 2012 International Conference on. IEEE, 2012: 1-6.
[2] Pan Z S, Chen S C, Hu G B, et al. Hybrid neural network and C4. 5 for misuse detection[C]//Machine Learning and Cybernetics, 2003 International Conference on. IEEE, 2003, 4: 2463-2467.
[3] Guo C, Ping Y, Liu N, et al. A two-level hybrid approach for intrusion detection[J]. Neurocomputing, 2016, 214: 391-400.
[4] Kim G, Lee S, Kim S. A novel hybrid intrusion detection method integrating anomaly detection with misuse detection [J]. Expert Systems with Applications, 2014, 41(4): 1690-1700.
[5] Jaiganesh V, Mangayarkarasi S, Sumathi P. Intrusion detection systems: A survey and analysis of classification techniques [J]. International Journal of Advanced Research in Computer and Communication Engineering, 2013, 2(4): 1629-1635
[6] Lee W, Stolfo S J. Data Mining Approaches for Intrusion Detection[C]//Usenix security. 1998.
[7] Keerthi S S, Gilbert E G. Convergence of a generalized SMO algorithm for SVM classifier design [J]. Machine Learning, 2002, 46(1-3): 351-360.
[8] Collobert R, Bengio S, Bengio Y. A parallel mixture of SVMs for very large scale problems[J]. Neural computation, 2002, 14(5): 1105-1114.
[9] Cataltepe Z, Ekmekci U, Cataltepe T, et al. Online feature selected semi-supervised decision trees for network intrusion detection[C]//Network Operations and Management Symposium (NOMS), 2016 IEEE/IFIP. IEEE, 2016: 1085-1088.
[10] Muniyandi A P, Rajeswari R, Rajaram R. Network anomaly detection by cascading k-Means clustering and C4. 5 decision tree algorithm [J]. Procedia Engineering, 2012, 30: 174-182.
[11] Patel J, Panchal K. Effective Intrusion Detection System using Data Mining Technique [J]. Journal of Emerging Technologies and Innovative Research, 2015, 2 (6).
[12] Al-Yaseen W L, Othman Z A, Nazri M Z A. Multi-level hybrid support vector machine and extreme learning machine based on modified K-means for intrusion detection system [J]. Expert Systems with Applications, 2017, 67: 296-303.
[13] Huang Tao, Liu Shenghui, Tan Yanna.Based on k-means clustering algorithm [J]. Computer Technology and Development, 2011, (07): 54-57 + 62.
[14] Tax D M J, Duin R P W. Support vector domain description [J]. Pattern recognition letters, 1999, 20(11): 1191-1199.
[15] Huang Qian, Wang Zhen, Wei Tao, Chen Yu. Real-time intrusion detection system based on One-class SVM [J]. Computer Engineering, 2006, (16): 127-129.
[16] Perdisci R, Gu G, Lee W. Using an ensemble of one-class svm classifiers to harden payload-based anomaly detection systems[C]//Data Mining, 2006. ICDM'06. Sixth International Conference on. IEEE, 2006: 488-498.
[17] Fisher W D, Camp T K, Krzhizhanovskaya V V. Anomaly detection in earth dam and levee passive seismic data using support vector machines and automatic feature selection [J]. Journal of Computational Science, 2017, 20: 143-153.
[18] Koshal J, Bag M. Cascading of C4. 5 decision tree and support vector machine for rule based intrusion detection system [J]. International Journal of Computer Network and Information Security, 2012, 4(8): 8.
[19] Muniyandi A P, Rajeswari R, Rajaram R. Network anomaly detection by cascading k-Means clustering and C4.5 decision tree algorithm [J]. Procedia Engineering, 2012, 30: 174-182.

IOP Conf. Series: Earth and Environmental Science **114** (2018) 012015 doi:10.1088/1755-1315/114/1/012015

An Evaluation Method of Equipment Reliability Configuration Management

Wei Wang[1], Weijia Feng[2],Wei Zhang[2] and Yuan Li[2]

1 Associate Professor, Beihang University, Beijing, China
2 Master Student, Beihang University, Beijing, China
E-mail: fwj_sprinkle9@163.com

Abstract. At present, many equipment development companies have been aware of the great significance of reliability of the equipment development. But, due to the lack of effective management evaluation method, it is very difficult for the equipment development company to manage its own reliability work. Evaluation method of equipment reliability configuration management is to determine the reliability management capabilities of equipment development company. Reliability is not only designed, but also managed to achieve. This paper evaluates the reliability management capabilities by reliability configuration capability maturity model(RCM-CMM) evaluation method.

1. Introduction

Modern equipment development companies have paid more and more attention to the reliability of the equipment development work and carried out the reliability program according to the GJB450A. However, in the equipment development process, it is always a problem how to evaluate the implementation of the reliability work. And because of this problem, it is difficult to achieve the reliability requirement of the equipment development. Therefore, it is necessary to propose a management method to enable the regularization of the reliability engineering management process and ultimately to ensure the realization of the reliability requirement.

This paper attempts to combine capability maturity model and reliability configuration management theory, establish evaluation and upgrade mechanism to help equipment development companies to understand and upgrade the reliability configuration management capability, which is contributed to the improvement in the overall equipment development reliability engineering management.

2. RCM-CMM Evaluation System

2.1. Overall structure

This paper will establish capability maturity model of equipment development reliability configuration management(RCM-CMM) to provide a method to evaluate and improve the capability of reliability management.

RCM-CMM is an application of CMM. This model has both CMM's generality and its own pertinence. As shown in the figure1, in this paper, the capability elements consist of RCM process field and RCM organization field. RCM process field includes main activities and procedures of CM; RCM organization field includes organization, personnel and technology.

IOP Conf. Series: Earth and Environmental Science **114** (2018) 012015 doi:10.1088/1755-1315/114/1/012015

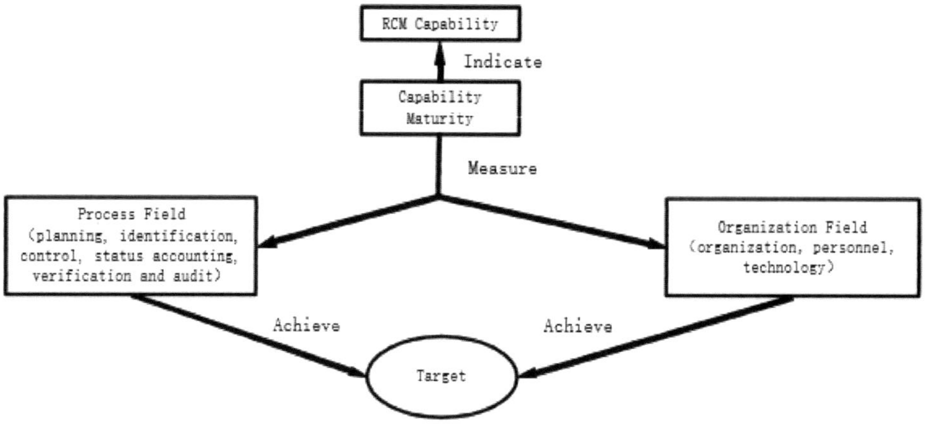

Figure 1. RCM management evaluation model.

2.2. Maturity level

Maturity level of RCM-CMM is shown as table1 referring to CMMI and SECM.

Table 1. Characteristic of RCM-CMM level.

Level	Name	Characteristic
1	Initial	Processes unpredictable, poorly controlled and reactive
2	Managed	Processes characterized for projects and is often reactive
3	Defined	Processes characterized for the organization and is proactive.(Projects tailor their processes from organization's standards)
4	Quantitatively Managed	Processes measured and controlled
5	Optimizing	Focus on process improvement

2.3. Design principles

In order to comprehensively, objectively and accurately reflect the capability of RCM-CM, the design of this evaluation system follows the following principles:

- Scientific principle;
- Integrity principle;
- Importance principle;
- The combination of qualitative and quantitative;
- Feasibility principle;
- Hierarchy principle.

3. RCM-CMM Evaluation Method

3.1. Screening and identifying evaluation index

This paper establishes an evaluation system of RCM-CMM at four levels which is shown as table2: the target level, the criterion level, the integrated parameter level and the basic parameter level. It decomposes the final target into several elementary parameters which can be estimated.

Table 2. Index of the evaluation system.

Target	Criterion	Integrated Parameter	Basic Parameter
RCM Capability	Process Field	RCM Plan	RCM Plan making and releasing
			RCM plan executing and controlling
		RC Identification	Choosing RC items
			Determining CM document requirement
			Determining RC identification program
			Setting and maintaining the baselines
			Interface management
		RC Control	Change control of closed circle
			Controlling deviation and tolerance
		RC Status Accounting	RC identification Status Accounting
			RC control Status Accounting
			RC verification and audit Status Accounting
		RC Verification and Audit	Internal Auditing the implementation of RCM plan
			Functional RC verification and audit
			Physical RC verification and audit
	Organization Field	RCM Organization	Determine RCM organization roles, rights and liabilities
			Coordinating inside and outside of RCM Organization
			Leader's attention of RCM organization
			Organizational management's strategies and objectives
			Organization culture and cohesion
		RCM Personnel	RCM Personnel overall investment
			Effective allocation of RCM personnel
			RCM personnel's master of professional knowledge
			RCM personnel's training and promotion
			RCM personnel's performance appraisal
		RCM Technology	Infrastructure such as networks
			RCM support tools
			Related specifications and certification of RCM
			Accumulation and delivery of knowledge and experience

IOP Conf. Series: Earth and Environmental Science **114** (2018) 012015 doi:10.1088/1755-1315/114/1/012015

3.2. Determining weight of index

Analyzing the importance of index of each level by using the analytic hierarchy process(AHP), and calculating the weight of the index.

(1) By pairwise comparing the index of same level, establish comparison matrix $A_n = (a_{ij})_{n\times n}$, where $a_{ij} > 0$, $a_{ii} = 1$, $a_{ji} = 1/a_{ij}$, $\mathrm{i,j} = 1,2,\ldots,\mathrm{n}$; One of the comparison matrix is shown as Table3;

Table 3. An example of comparison matrix.

	RCM Plan	RC Identification	RC Control	RC Status Accounting	RC Verification and Audit
RCM Plan	a_{11}	a_{12}	a_{13}	a_{14}	a_{15}
RC Identification	a_{21}	a_{22}	a_{23}	a_{24}	a_{25}
RC Control	a_{31}	a_{32}	a_{33}	a_{34}	a_{35}
RC Status Accounting	a_{41}	a_{42}	a_{43}	a_{44}	a_{45}
RC Verification and Audit	a_{51}	a_{52}	a_{53}	a_{54}	a_{55}

(2) Letting $|A_n - \lambda E| = 0$, Calculate the maximum eigenvalue of matrix A_n;
(3) Using the eigenvalue to test the consistency according to AHP method. The result is acceptable if $\mathrm{CR} \le 0.1$ or $\lambda = N, \mathrm{CR} = 0$;
(4) Calculate weighting vector $\mathrm{W} = \{w_1, w_2, \ldots, w_n\}$ by $A_n W = \lambda_{max}\mathrm{W}$;
(5) The weight collection of index is shown as follows.
Criterion level:$\mathrm{A} = \{a_1, a_2\}$;
Integrated parameter level:
Process field:$\mathrm{A}_1 = \{a_{11}, a_{12}, a_{13}, a_{14}, a_{15}\}$;
Organization field:$\mathrm{A}_2 = \{a_{21}, a_{22}, a_{23}\}$;
Basic parameter level:
RCM Plan:$\mathrm{A}_{11} = \{a_{111}, a_{112}\}$;
RC Identification: $\mathrm{A}_{12} = \{a_{111}, a_{122}, a_{123}, a_{124}, a_{125}\}$;
RC Control: $\mathrm{A}_{13} = \{a_{131}, a_{132}\}$;
RC Status Accounting: $\mathrm{A}_{14} = \{a_{141}, a_{142}, a_{143}\}$;
RC Verification and Audit: $\mathrm{A}_{15} = \{a_{151}, a_{152}, a_{153}\}$;
RCM Organization: $\mathrm{A}_{21} = \{a_{211}, a_{212}, a_{213}, a_{214}, a_{215}\}$;
RCM Personnel: $\mathrm{A}_{22} = \{a_{221}, a_{222}, a_{223}, a_{224}, a_{225}\}$;
RCM Technology: $\mathrm{A}_{23} = \{a_{231}, a_{232}, a_{233}, a_{234}\}$.

3.3. Establishing scoring criteria of index

This paper uses experts scoring method to carry out evaluation. The scoring criteria are shown as table4.

Table4. Scoring criteria of the evaluation of index.

Integrated Parameter	Basic Parameter	Maturity Level				
		Initial	Managed	Defined	Quantitatively Managed	Optimizing
RCM Plan	RCM Plan making and releasing	None or disorder	project-level management	Standardization of organization and systematization	Quantitatively management	Continuous optimization
…	…	…	…	…	…	…
RCM organization	Determine RCM organization roles, rights and liabilities	Very bad	Bad	Average	Good	Excellent
…	…	…	…	…	…	…

3.4. Maturity Calculating

(1) This paper uses a membership function whose span is [0,1] to quantify index and use experts scores to get fuzzy evaluating matrix with single factor.

Degree of membership r_{ijk}^{m} indicates the membership degree of mth level v_m of factor U_{ijk}.

$$r_{ijk}^{m} = \frac{d_{ijk}^{m}}{d} \tag{1}$$

Where d is the number of experts, d_{ijk}^{m} is the number of experts whose score is v_m.

(2) Fuzzy evaluating matrix:

$$\text{RCM Plan:}R_{11} = \begin{bmatrix} r_{111}^{1} & r_{111}^{2} & r_{111}^{3} & r_{121}^{4} & r_{121}^{5} \\ r_{112}^{2} & r_{112}^{2} & r_{112}^{3} & r_{112}^{4} & r_{112}^{5} \end{bmatrix}$$

$$\text{RC identification:}R_{12} = \begin{bmatrix} r_{121}^{1} & r_{121}^{2} & r_{121}^{3} & r_{121}^{4} & r_{121}^{5} \\ r_{122}^{1} & r_{122}^{2} & r_{122}^{3} & r_{122}^{4} & r_{122}^{5} \\ r_{123}^{1} & r_{123}^{2} & r_{123}^{3} & r_{123}^{4} & r_{123}^{5} \\ r_{124}^{1} & r_{124}^{2} & r_{124}^{3} & r_{124}^{4} & r_{124}^{5} \\ r_{125}^{1} & r_{125}^{2} & r_{125}^{3} & r_{125}^{4} & r_{125}^{5} \end{bmatrix}$$

……

(3) Maturity evaluation vector of single factor:

$$\text{RCM Plan:}B_{11} = A_{11} \cdot R_{11} = (a_{111}, a_{112}) * \begin{bmatrix} r_{111}^{1} & r_{111}^{2} & r_{111}^{3} & r_{121}^{4} & r_{121}^{5} \\ r_{112}^{2} & r_{112}^{2} & r_{112}^{3} & r_{112}^{4} & r_{112}^{5} \end{bmatrix}$$

$$\text{RC identification:}R_{12} = A_{12} \cdot R_{12} = (a_{111}, a_{122}, a_{123}, a_{124}, a_{125}) * \begin{bmatrix} r_{121}^{1} & r_{121}^{2} & r_{121}^{3} & r_{121}^{4} & r_{121}^{5} \\ r_{122}^{1} & r_{122}^{2} & r_{122}^{3} & r_{122}^{4} & r_{122}^{5} \\ r_{123}^{1} & r_{123}^{2} & r_{123}^{3} & r_{123}^{4} & r_{123}^{5} \\ r_{124}^{1} & r_{124}^{2} & r_{124}^{3} & r_{124}^{4} & r_{124}^{5} \\ r_{125}^{1} & r_{125}^{2} & r_{125}^{3} & r_{125}^{4} & r_{125}^{5} \end{bmatrix}$$

……

(4) Maturity evaluation vector of multiple factors:

Process field: $P = A_1 \cdot B_1 = A_1 * \begin{bmatrix} B_{11} \\ B_{12} \\ B_{13} \\ B_{14} \\ B_{15} \end{bmatrix}$

Organization field: $Q = A_2 \cdot B_2 = A_2 * \begin{bmatrix} B_{21} \\ B_{22} \\ B_{23} \end{bmatrix}$

(5) Synthetic evaluation vector: $W = A * \begin{bmatrix} P \\ Q \end{bmatrix}$

(6) Calculation of comprehensive evaluation value: $M = W \cdot F^T = W * (1,2,3,4,5)^T$

3.5. Result Analysis

According to M, we can know the level of RCM-CM. Then the description of this level can be found in table1. This is a scientific conclusion. It can be a basis for decision making about equipment checking and acceptance.

4. Conclusion

In this paper, a method to evaluation the RCM-CM based on AHP method is proposed.

First, this paper defines RCM-CMM. Then it presents an evaluation system for the evaluation of equipment development reliability. The evaluation result can not only reflect the level of RCM-CM scientifically, but also provide the basis of acceptance decision for consumers.

An open problem is still precision, consistency and efficiency of the method. Future work will be focused on the development of a sophisticated method of deriving an optimized high-precision result.

5. References

[1] Xiwei Huang. Analysis and practice of CMMI, Beijing: Posts and Telecommunications Press, (2004)

[2] Enterprise Process Improvement Collaboration. EPI capability maturity model for systems engineering (SCMM). Version 1.1. EPIC, (1996)

[3] T.F. Burgess, D. Mc Kee, C. Kidd. Configuration management in the aerospace industry: a review of industry practice. International Journal of Operations & Production Management,25(3):290-301, (2005)

[4] Lyu M R. Software reliability engineering: A roadmap[C]//2007 Future of Software Engineering. IEEE Computer Society, 153-170, (2007)

[5] Guo Y, Deng Y, Ren J. Application of Analytic Hierarchy Process(AHP) Method in Comprehensive Assessment of Surface Water Quality[J]. Journal of Lanzhou Jiaotong University, (2006)

IOP Conf. Series: Earth and Environmental Science **114** (2018) 012016 doi:10.1088/1755-1315/114/1/012016

Research and analysis on response characteristics of bracket-line coupling system under wind load

Zhao Jiayu[1] and Sun Qing[2]

1 Department of Civil Engineering, Xi'an Jiaotong University, China
E-mail: 137785151@qq.com
2 Department of Civil Engineering, Xi'an Jiaotong University, China
E-mail: sunq@mail.xjtu.edu.cn

Abstract. In this paper, a three-dimensional finite element model of bracket-line coupling system is established based on ANSYS software. Using the wind velocity time series which is generated by MATLAB as a power input, by comparing and analyzing the influence of different wind speeds and different wind attack angles, it is found that when 0 degree wind acts on the structure, wires have a certain damping effect in the bracket-line coupling system and at the same wind speed, the 90 degree direction is the most unfavorable wind direction for the whole structure according to the three kinds of angle wind calculated at present. In the bracket-line coupling system, the bracket structure is more sensitive to the increase of wind speed while the conductors are more sensitive to the change of wind attack angle.

1. Introduction

In modern industry, composite materials are artificial, synthetic, two-phase or multiphase materials, usually one phase reinforced material, and the other as matrix. The most commonly used composite material in industry is the fiber reinforced plastic (FRP). Because this composite material can be combined with different materials to play a role, it has the advantages of fatigue resistance, corrosion resistance, electromagnetic shielding and long service life and widely used in aviation, aerospace and other fields which requiring light and high-strength structures [1].

FRP bracket-wire coupling system has the characteristics of high span and large span, its structure is complex, sensitive to the wind, very easy to cause vibration under the action of wind load, causing some damage and fatigue failure of components, and even cause the collapse of the whole structure, thus bring about serious consequence [2]. The wind-induced response of the FRP bracket-wire coupling system is mainly caused by the fluctuating wind, therefore, the accurate simulation of wind load, especially the simulation of random wind with random characteristics, is the key to ensure the safety and reliability of structural system under wind load. Domestic and foreign scholars have done a lot of research on the characteristics of fluctuating wind, and proposed two important characteristics of fluctuating wind—fluctuating wind velocity spectra and empirical formula of spatial correlation [3-4].

In this paper, firstly, using finite element software to establish the overall structure and using numerical simulation method to simulate the wind speed. Then, under the wind speed of 30m/s, the wind-induced response of the structure at three different angles (0 degrees, 60 degrees, 90 degrees) is analyzed by ANSYS. At last, the response of bracket-line coupling system is calculated under the action of 45m/s wind speed.

2. Finite element model

According to the design drawings, the largest span of brackets is 61.8m, the maximum height is 30.8m, four-span brackets connect with 300 wires and the ZJ-4 bracket is connected to the ground anchor through 30 FRP tie bars. The total length of the whole structure system is 198m, the span is 61.8m, and the height is 30.8m. The support beam is mainly made of Q345 steel pipe, and the support column is FRP pipe connected with steel joint.

Taking into account the actual force situation, the BEAM189 unit is selected in the ANSYS to simulate the main force components of the support structure and the LINK10 unit is used to simulate the traverse. Finally, the three span wires and the FRP tie rod are assembled and connected with the four brackets, and the finite element model is shown in figure 1.

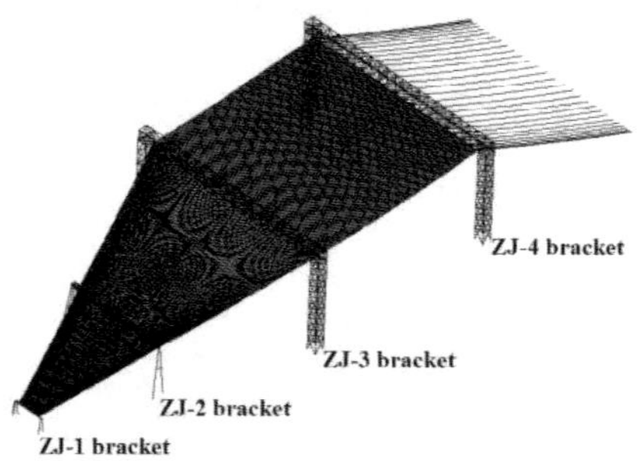

Figure 1. Finite element model of the structure

After the finite element model is established, modal analysis of the four supports is carried out respectively, and the natural frequencies of the structures are shown in Table 1. Figure 2 shows the first three modes of the ZJ-3 bracket.

Table 1. The first five modal frequencies of each bracket

Order of modes	No.ZJ-1 bracket	No.ZJ-2 bracket	No.ZJ-3 bracket	No.ZJ-4 bracket
1	10.221 Hz	1.7428 Hz	1.1624 Hz	1.1308 Hz
2	12.002 Hz	2.1605 Hz	1.6530 Hz	1.6043 Hz
3	19.088 Hz	3.5025 Hz	1.7416 Hz	1.7467 Hz
4	25.846 Hz	4.2863 Hz	2.9711 Hz	2.8905 Hz
5	42.090 Hz	4.5686 Hz	3.3161 Hz	3.8207 Hz

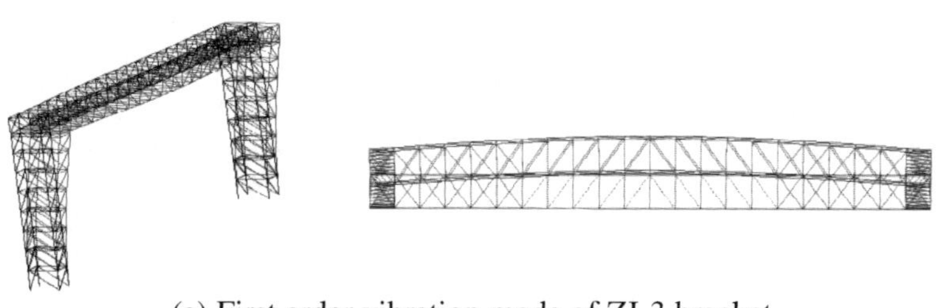

(a) First order vibration mode of ZJ-3 bracket

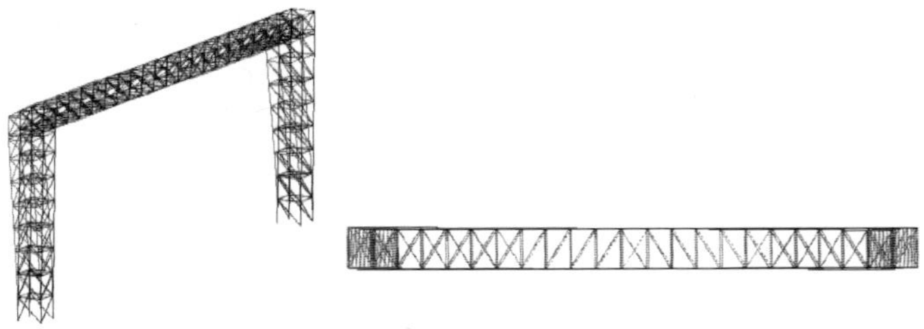

(b) Second order vibration mode of ZJ-3 bracket

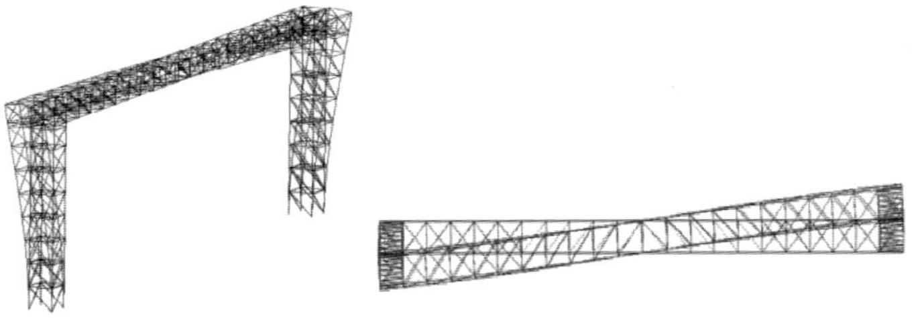

(c) Third order vibration mode of ZJ-3 bracket

Figure 2. First three modes of the ZJ-3 bracket

3. Numerical simulation of wind field

According to the basic properties of fluctuating wind, Simiu spectrum and linear autoregressive filter method [5-7], a program of Simiu spectrum-time curve has been compiled in the MATLAB, and the fluctuating wind speed at different heights has been obtained. In this paper, the fitting curves of the simulated spectrum and the target spectrum at two different heights are given.

It can be seen from Fig.3 that the simulated wind spectrum of each layer agrees well with the target spectrum, so the simulated wind can be used to analyze the wind-induced response.

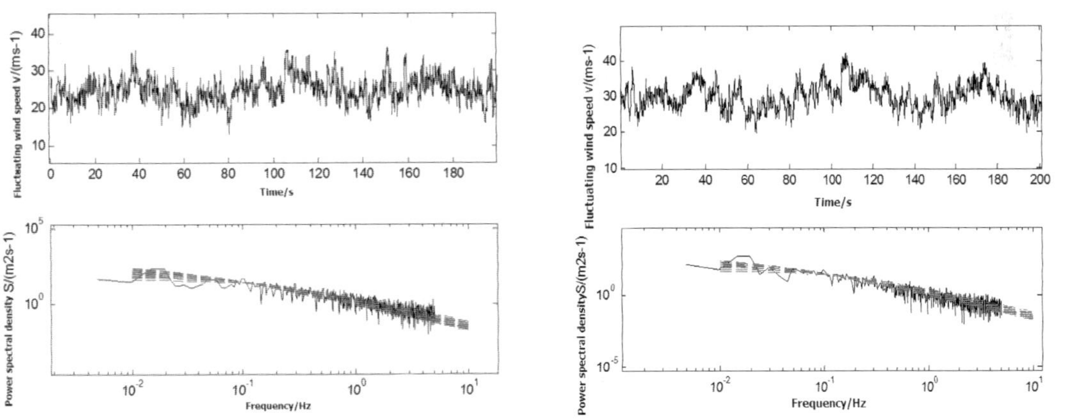

（a）Fluctuating wind speed-time curve at hight of 3.75m （b）Fluctuating wind speed-time curve at height of 11.25m

Figure 3. Fluctuating wind speed-time curve

4. Wind-induced response analysis of bracket-line coupling system

4.1. Loaning mode

The structure is layered according to the height and applied with corresponding wind loads, take the No.ZJ-3 bracket as an example. The column is divided into 5 layers and the wind load of each layer is added to the 4 nodes and the beam is divided into 11 sections to apply wind load. The loading mode is shown in figure 4, where the arrow refers to the position of the loading point.

The application of wind load on conductors: Along the longitudinal direction of the structure, the first-span conductors are divided into 5m sections to be loaded, the second-span conductors are divided into 5m sections to be loaded, the third-span conductors are divided into 15m sections to be loaded and the fourth-span conductors are divided into 5m sections to be loaded. The details are shown in figure 5.

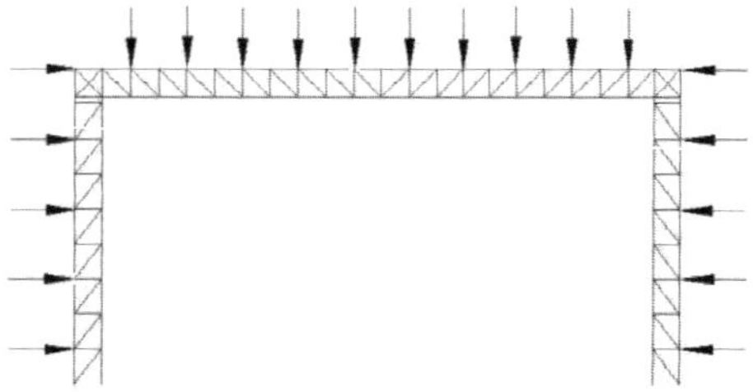

Figure 4. Loading mode of brackets

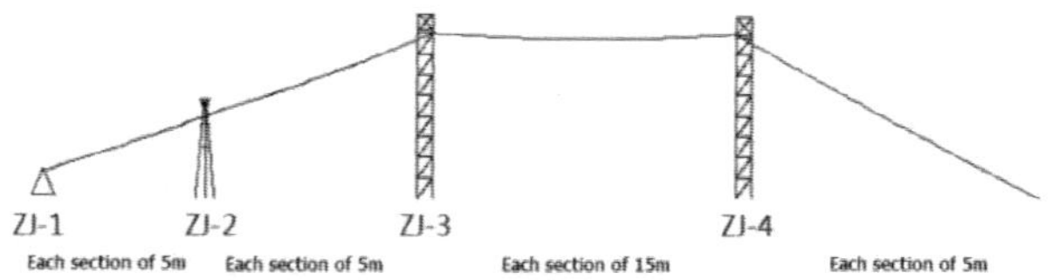

Figure 5. Loading mode of wires

4.2. Time response curve under wind load excitation (the wind speed is 30m/s)

Considering the damping ratio (0.03), the wind-induced vibration response of the structure is analyzed. Figure 6 is the displacement-time curve of the top of No.ZJ-3 bracket, figure 7 is Displacement-time curve of third span conductors along vertical direction of the wire, figure 8 is the Vertical displacement-time curve of third span conductors and figure 9 is the axial force-time curve of third span conductors.

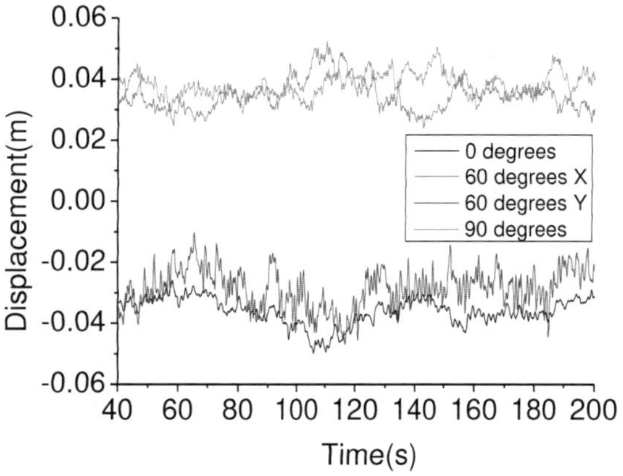

Figure 6. Displacement-time curve of the top of No.ZJ-3 bracket

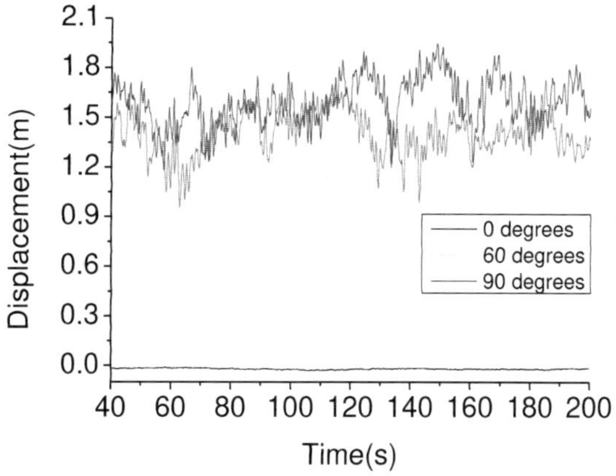

Figure 7. Displacement-time curve of third span conductors along vertical direction of the wire

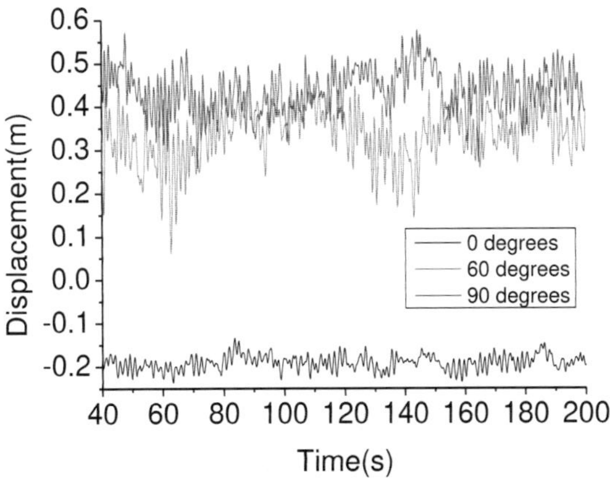

Figure 8. Vertical displacement-time curve of third span conductors

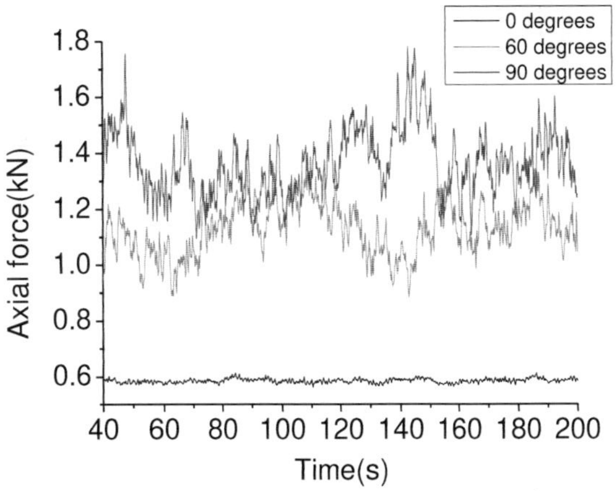

Figure 9. Axial force-time curve of third span wires

It is not difficult to see from the diagram that the effect of wind at different angles on the structure is different. When the 90 degree wind acts, the brackets have the largest displacement response, while the 0 degree wind has the least influence on the brackets. This is mainly because: In the bracket-line coupling system, When the 0 degree wind acts on the structure, Wind-shield area of the wire is small, as a result, it is less affected by wind load. What is more, the wire itself has a certain weight, therefore, wires have a certain damping effect in the whole structure.

Table 2 is the maximum displacement of the top of each bracket along the wind direction under the influence of wind at different angles. (When 60 degrees wind acts, there are two directions of displacement).

Table 2 Maximum displacement of each bracket

	No.ZJ-1 bracket	No.ZJ-2 bracket	No.ZJ-3 bracket	No.ZJ-4 bracket
0 degrees of wind	0.017m	0.028m	0.050m	0.025m
90 degrees of wind	0.002m	0.087m	0.052m	0.035m
60 degrees of wind	0.009m 0.017m	0.011m 0.074m	0.047m 0.052m	0.020m 0.035m

From the table we can find that under the influence of 30m/s wind speed, different angles of the wind on each bracket will have a certain impact, but at any angle of the wind, the top displacement response of the ZJ-2 and ZJ-3 brackets is the largest in the four supports. This is mainly due to the fact that the two supports are in adjacent positions, and the height difference is large. Therefore, when the wind load is applied, a large unbalance force is produced between the two supports, which make the displacement response more obvious.

Table 3 shows the maximum tension at the mid-span of conductors under the influence of wind at different angles

Table 3.Maximum tension at the mid-span of conductors

	First-span wires	Second-span wires	Third-span wires	Fourth-span wires
0 degrees of wind	1.023 kN	0.872 kN	0.611 kN	4.432 kN
90 degrees of wind	1.591 kN	1.646 kN	1.781 kN	16.720 kN
60 degrees of wind	1.467 kN	1.434 kN	1.387 kN	12.373 kN

Comparative analysis of wire tension in the maximum value under the effect of different wind angle, we can see that no matter what kind of wind angle acts in the structure, the fourth wire has the largest value of tension. This is mainly due to the fourth wire compared to other wires, has the characteristics of large diameter and the end is directly connected with the anchor. This allows it to play a stabilizing role in the overall structure and therefore bears greater power.

4.3. Spectral response under wind excitation (the wind speed is 30m/s, 90 degrees of wind)

In this paper, the displacement spectrum and acceleration spectrum of the support and the displacement spectrum of the wire are obtained by the Fast Fourier Transform (FFT). They are shown in figure 10, figure 11 and figure 12. Figure 13 also gives the amplitude-frequency curve of the wind load.

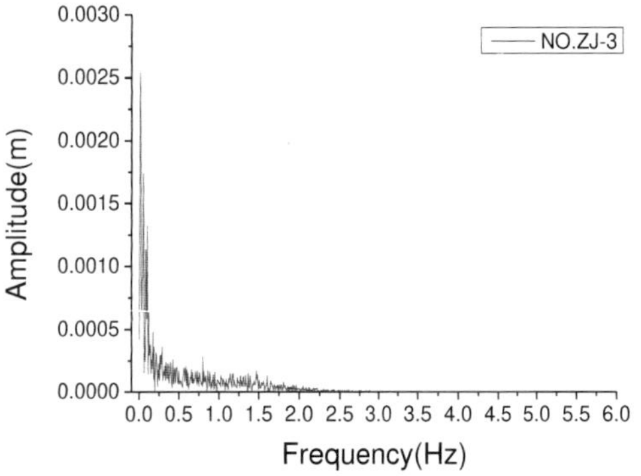

Figure 10. Displacement spectrum of support

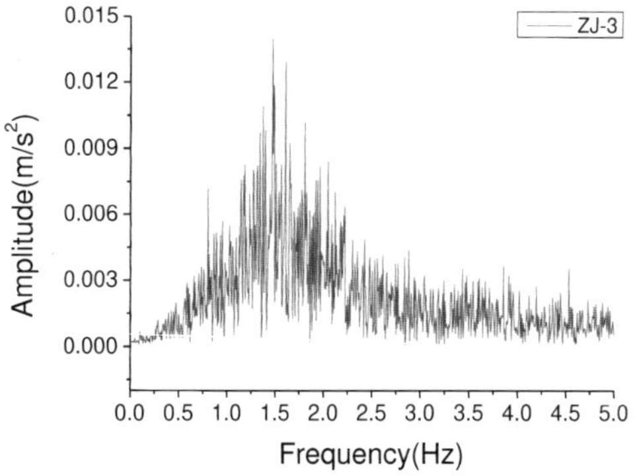

Figure 11. Acceleration spectrum of support

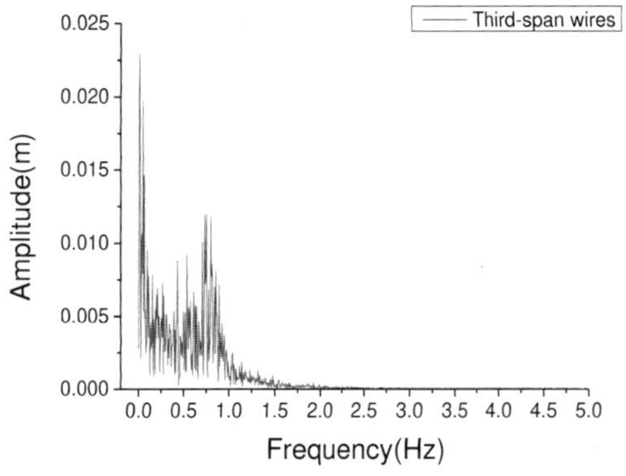

Figure 12. Displacement spectrum of wires

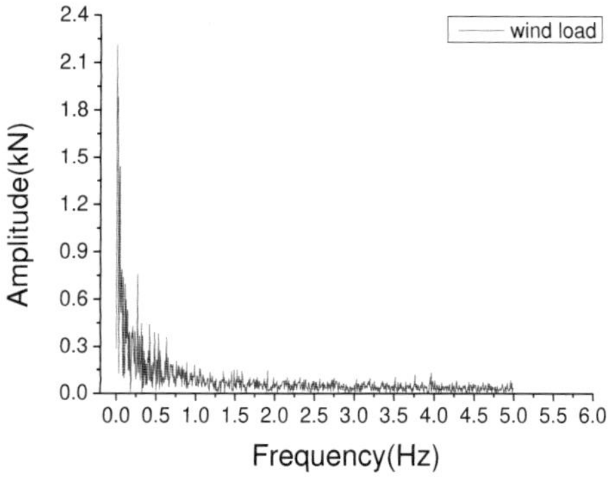

Figure 13. Amplitude-frequency curve of wind load

From above figures, it can be found that the displacement and acceleration spectrum of the top of bracket are mainly concentrated in the low frequency region and in the top displacement response of the brackets, the displacement response at low frequency is similar to that of wires. This shows that the influence of the conductors on the top displacement of bracket is obvious in the bracket-line coupling system. Comparing the spectrum curves of wind load and the displacement of support, it is easy to see that the two curves have great similarity, and all of them are mainly concentrated in the low frequency region. This shows that the displacement response of the support is greatly affected by wind load.

4.4. Comparison of structure response at different wind speeds (the wind speeds are 30m/s and 45m/s)

In the 90 degree wind condition, structural response in 30m/s wind and 45m/s wind speed are calculated respectively. Figure14, 15, 16 and 17 show the displacement-time curve of the top of No.ZJ-3 bracket, the acceleration-time curve of No.ZJ-3 bracket, the displacement-time curve and the axial force-time curve of third span wires.

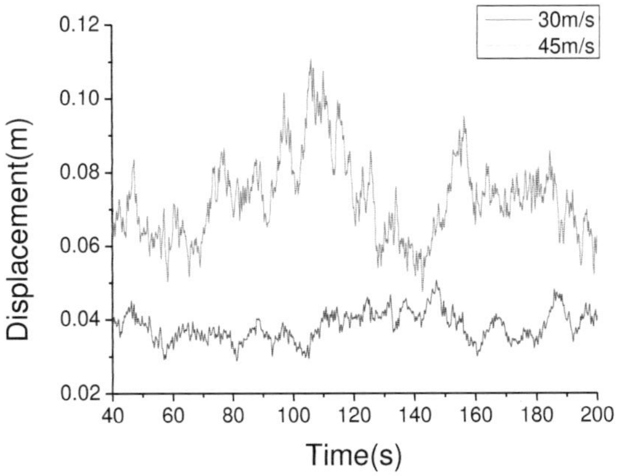

Figure 14. Displacement-time curve of the top of No.ZJ-3 bracket

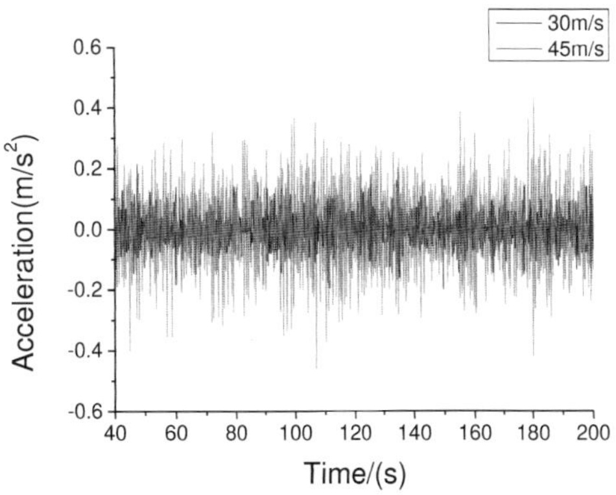

Figure 15. Acceleration-time curve of No.ZJ-3 bracket

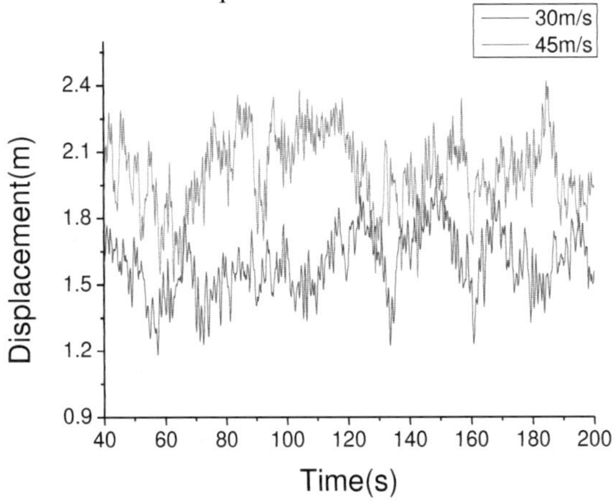

Figure 16. Displacement-time curve of third span wires

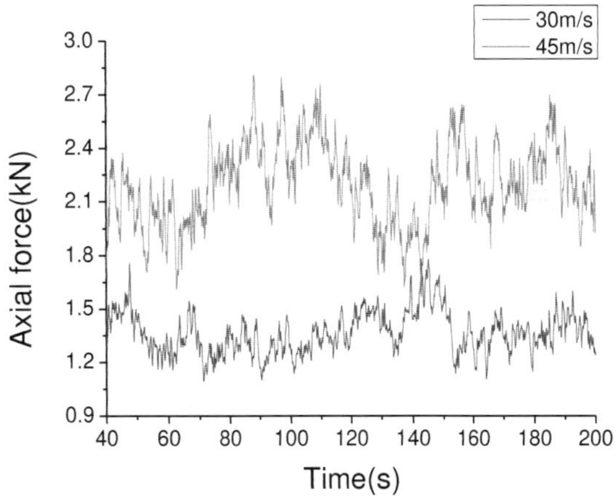

Figure 17.Axial force-time curve of third span wires

As we can see from the picture above: At the same angle of wind (90 degree wind), after the wind speed increased by 1.5 times, both the response of brackets and wires have increased obviously. According to the displacement-time curve of No.ZJ-3 bracket, the displacement-time curve of 45m/s wind is completely above the curve of the 30m/s wind and its maximum value is about 2 times than that of 30m/s wind acts, what is more, the acceleration of the top is generally larger than that of the 30m/s wind. For the wires, the displacement response is also increased when the 45m/s wind acts, but its displacement-time curve partly overlaps with the curve of 30m/s wind. The maximum displacement is about 1.3 times than that of 30m/s wind acts and the maximum tension is about 1.3 times than that of 30m/s wind acts.

5. Conclusions

In this paper, the following conclusions are obtained by simulating the fluctuating wind speed time series and analyzing the wind-induced vibration response of the FRP bracket-line coupling system:

(1) In the bracket-line coupling system, When the 0 degree wind acts on the structure, Windshield area of the wire is small, as a result, wires have a certain damping effect in the whole structure, so in the case of 30m/s wind, When the 90 degree wind acts, the brackets have the largest displacement response, while the 0 degree wind has the least influence on the brackets.

(2) Under the same wind speed, the influence of the 0 degree wind on the conductors is relatively small, and the displacement and tension response of the conductor is more obvious when the 90 degree wind acts.

(3) Considering the effects of wind on the brackets and the conductors at three different angles, we can see that in the case of 30m/s wind, the wind in the 90 degree direction is the most unfavorable wind direction for the whole structure according to the three kinds of angle wind calculated at present.
(4) The displacement and acceleration spectrum of the top of brackets are mainly concentrated in the low frequency region and the displacement response of brackets at low frequency is similar to that of wires. This shows that the influence of the conductors on the top displacement of brackets is obvious in the bracket-line coupling system.
(5) By comparing and analyzing the influence of different wind speeds and different wind attack angles, it is found that in the bracket-line coupling system, the bracket structure is more sensitive to the increase of wind speed while the conductors are more sensitive to the change of wind attack angle.

6. References

[1] Zhao J, Cai G, Cui L, et al 2017 Deterioration of Basic Properties of the Materials in FRP-Strengthening RC Structures under Ultraviolet Exposure *J Polymers*. 9(9):402.

[2] Hui Y, Tamura Y, Yoshida A, et al 2013 Pressure and flow field investigation of interference effects on external pressures between high-rise buildings *J Wind Engineering & Industrial Aerodynamics*. 115(115):150-161.

[3] Tanaka H, Tamura Y, Ohtake K, et al. 2012 Experimental investigation of aerodynamic forces and wind pressures acting on tall buildings with various unconventional configurations *J Wind Engineering & Industrial Aerodynamics*. s 107–108(8):179-191.

[4] Wang D, Tse T K T, Zhou Y, et al 2015 Structural performance and cost analysis of wind-induced vibration control schemes for a real super-tall building *J Structure & Infrastructure Engineering*. 11(8):990-1011.

[5] Hino M, Hasebe M 1984 Identification and prediction of nonlinear hydrologic systems by the filter-separation autoregressive (AR) method: Extension to hourly hydrologic data *J Hydrolog*. 68(1):181-210.

[6] ÜllarRannik, Vesala T 1999 Autoregressive filtering versus linear detrending in estimation of fluxes by the eddy covariance method *J Boundary-Layer Meteorology*. 91(2):259-280.

[7] Masseran N. 2016 Modeling the fluctuations of wind speed data by considering their mean and volatility effects *J Renewable & Sustainable Energy Reviews*. 54:777-784.

Experimental study on the connection property of full-scale composite member

Cao Panpan[1] and Sun Qing[2]

1 Department of Civil Engineering, Xi'an Jiaotong University, China
E-mail: 574093483@qq.com
2 Department of Civil Engineering, Xi'an Jiaotong University, China
E-mail: sunq@mail.xjtu.edu.cn

Abstract. The excellent properties of composite result in its increasingly application in electric power construction, however there are less experimental studies on full-scale composite member connection property. Full-scale experiments of the connection property between E-glass fiber/epoxy reinforced polymer member and steel casing in practical engineering have been conducted. Based on the axial compression test of the designed specimens, the failure process and failure characteristics were observed, the load-displacement curves and strain distribution of the specimens were obtained. The finite element analysis was used to get the tensile connection strength of the component. The connection property of the components was analyzed to provide basis of the casing connection of GFRP application in practical engineering.

1. Introduction

Because of its high specific strength, good fatigue performance and strong design ability, fiber composite material has been gradually popularized in the field of structural engineering [1-3], especially in power construction in recent years [4-6]. Compared with other types of composite materials, such as carbon fiber reinforced plastic (CFRP), aramid fiber reinforced plastic (AFRP), glass fiber reinforced polymer (GFRP) is of better toughness (breaking elongation of 2.5%~3%) and economy, which results in its broader engineering application as a force component [7-9]. However, the application in structural construction still requires the development of reliable and convenient connection approaches, especially for tubular sections [10].

At present, there is less research in structural engineering field about the GFRP component, and the research is mainly focused on the material properties and the performance of components under small size. As for the composite component, the connection occupies a very important position, which is similar to the steel structural member. In this paper, the extruding test of bonded steel sleeve connection for full-scale GFRP component is carried out, the connection property between GFRP and steel sleeve is studied as well. Besides that, finite element (FE) analysis of the component under tension force is also conducted to estimate the joint capacity. The research could provide the design basis for engineering application of glass fiber composite materials.

2. Experiment condition

IOP Conf. Series: Earth and Environmental Science **114** (2018) 012017 doi:10.1088/1755-1315/114/1/012017

2.1. specimen details
The specimen was made up of two components. The epoxy resin was used as the matrix, and the glass fiber was used as the strengthening material. The form of the thin wall ring was adopted, and the forming process was winding forming. Between the sleeve and the GFRP pipe, the high strength adhesive was utilized to fix. In addition, the sleeve was grooving in the interface so as to improve the performance of the connection. The specimen size and arrangement of measuring points are shown in figure 1.

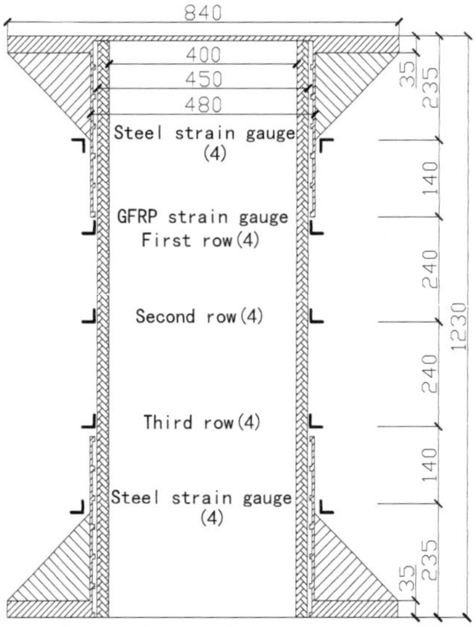

Figure1 The specimen size and arrangement of measuring points

2.2. Load scheme
The YAW-10000E microcomputer controlled electro-hydraulic servo pressure testing machine was used as the loading device. Compression test adopted load control before the pressure reached 4000kN. The loading speed was 5kN/s. Each level load was 100kN, and the loading interval of each level load was 30s. When the pressure was greater than 4000kN, the displacement control was adopted until the specimen lost bearing capacity. The loading speed was 2mm/min, each level of displacement was 1mm, and the loading interval of each level load was also 30s.

The extruding test needed an appropriate board at the end of the specimen, which is necessary for the GFRP pipe to be extruded. The bottom of the specimen and the board are shown in figure 2.

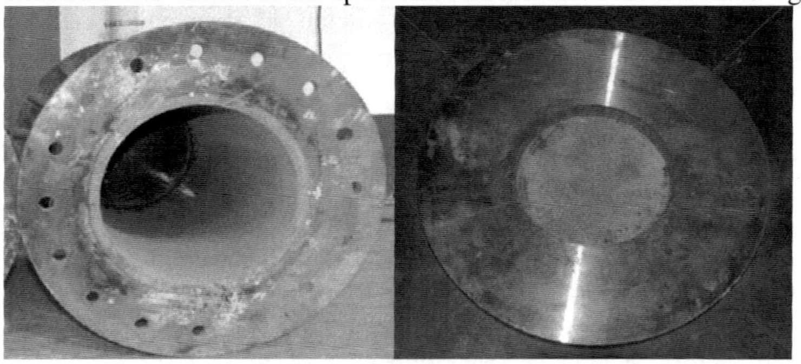

Figure 2 The bottom of the specimen and the board

3. Experimental phenomenon

The phenomenon observed in the course of the experiment is shown in the Fig. 3. There was no phenomenon but some sounds appearing on the specimens during the loading process. Set specimen PH-1 as an example to demonstrate how the figure shows. When the loading force was 100kN, the specimen made a faint sound. When the loading force was 1610kN, the specimen made a medium sound. When the loading force was 2350kN and 2610kN, the specimen both made a larger sound. When the loading force was 3750kN, the specimen made a medium sound. When the loading force was 4500kN, the specimen made a faint sound. When the loading force was 4950kN, the specimen was broken with a very big sound. Through further observation, the connection of specimen was broken before the GFRP pipe was destroyed by sudden impact. As we can see from the three curves, the concentration of sounds occurred between 1500kN and 3000kN. It can be inferred that the adhesive part was beginning to break.

The failure of the specimen is shown in Fig.4. The GFRP pipes of the three specimens were all extruded, PH-1 was pressed out 12mm, PH-2 was 20mm and PH-3 was 2mm. In addition, the GFRP pipes were damaged, which proved that the strength of the connection was equal to the damage strength of the material under the compression experiment.

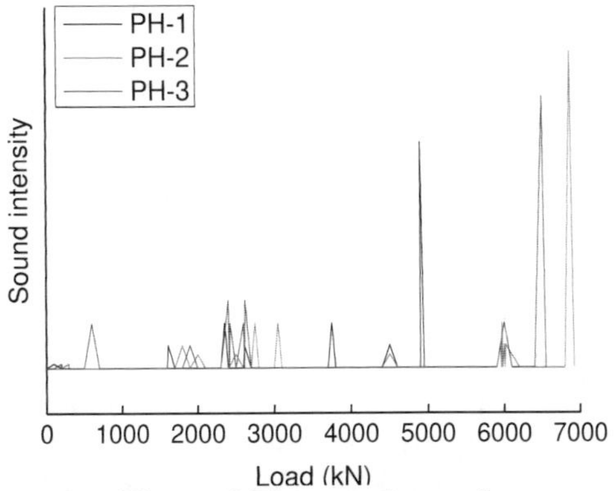

Figure 3 The sound record

(a) PH-1 specimen

(b) PH-2 specimen

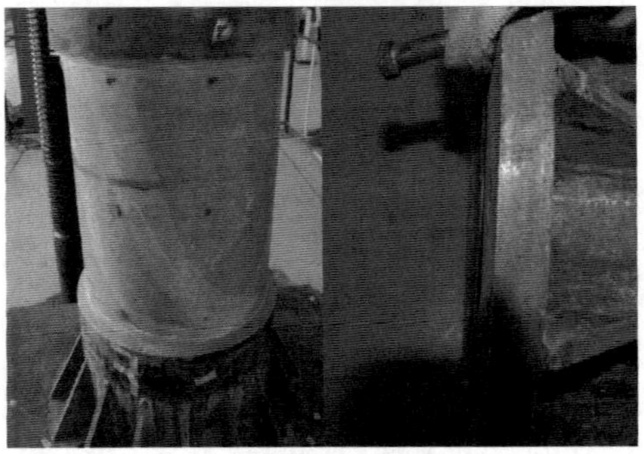

(c) PH-3 specimen

Figure 4 Destroyed specimens

4. Test result and analysis

The load-displacement curve of specimen is shown in Fig.5. As can be seen from the load-displacement curve of the three specimens, the curve can be divided into three phases, the linear segment, the broken line segment and the descending segment. There is a turning point between the linear and the broken line segment for each specimen, and when the load is greater than this value, the slope of the curve drops. When the load reaches limit finally, load-displacement curve falls sharply and specimen is damaged.

IOP Conf. Series: Earth and Environmental Science **114** (2018) 012017 doi:10.1088/1755-1315/114/1/012017

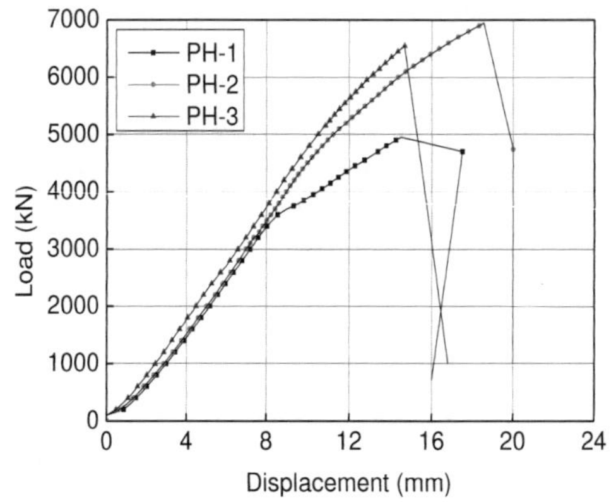

Figure5 The load-displacement curve

As we can see from the curve slope, the component of the overall stiffness is larger at the beginning, after a certain load, the component of the overall stiffness decreases. The load-displacement curve is basically linear, which shows that the component is brittle and the destruction is sudden without any sign.

The performance of the specimen is analyzed and summarized in table 1. The critical load of the specimens is between 4950kN and 6950kN, the mean value is 6150kN, and the mean of compressive strength is 184.34Mpa. The standard deviation of the test data of the ultimate bearing capacity of the component is 1058.3kN, which shows that the dispersion of connection performance is quite large.

If the adhesive area is supposed to be equivalent to the projection of the GFRP pipe to calculate the strength, the shear strength of the adhesive is 11.61MPa.

Table 1: Test specimen bearing capacity and test specimen adhesive strength

specimen	section size (mm•mm)	bearing capacity (kN)	average bearing capacity (kN)	standard deviation	equivalent adhesive area (mm2)	adhesive strength (Mpa)	average adhesive strength (Mpa)
PH-1		4950				9.34	
PH-2	Φ450•25	6950	6150	1058.30	529875	13.12	11.61
PH-2		6550				12.36	

The longitudinal maximum strain of the PH-1 GFRP pipe is -8524 με, the PH-2 is -9318 με and the PH-3 is -10030 με. The GFRP pipe is in the elastic stage from applying load to the failure, the load and strain remains generally a linear relationship. The strain increases with the increase of load. The load-longitudinal strain curve of the PH-3 is shown in the figure6. For the steel sleeve, far from the yield value, the longitudinal strain values are less than 800 με. Because the steel sleeve and the GFRP pipe are bonded together, the adhesive damages in the process of loading test and then sleeve will appear stress redistribution, stress situation was complex.

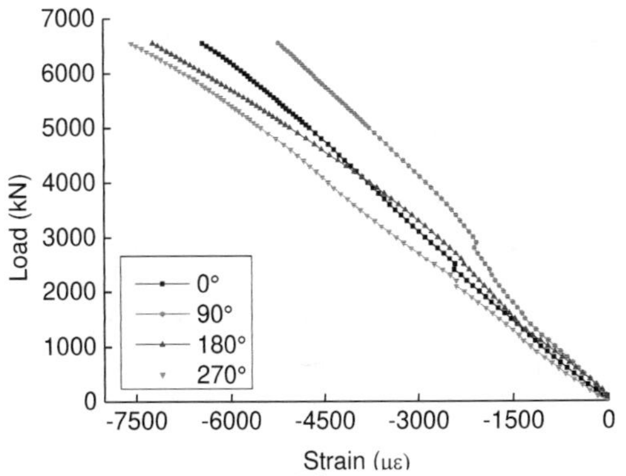

(a) First row

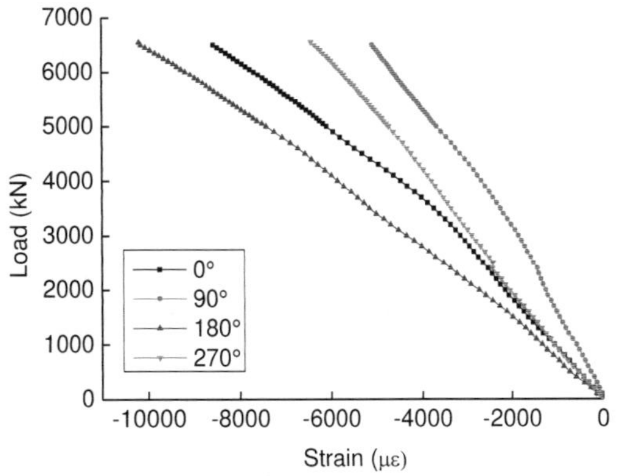

(b) Second row

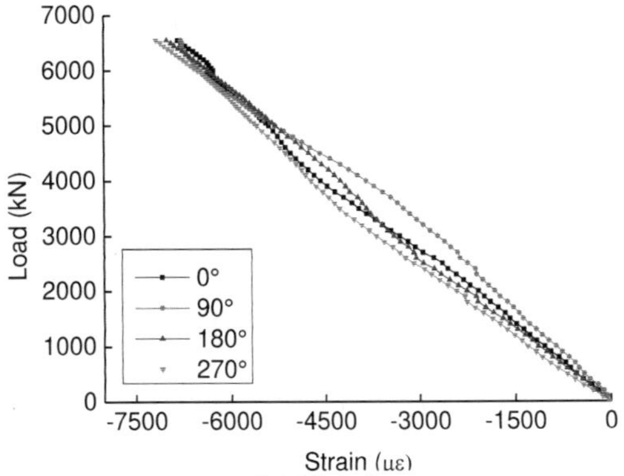

(c) Third row

Figure6. The GFRP load-longitudinal strain curve of PH-3

5. Finite element analysis

In this paper, the tensile strength of the same specimen is obtained by the finite element simulation. Before that the compression simulation of the same model was carried out, and the correctness of the finite element model is verified by comparing with the experimental results. The material test had been carried out before the production of specimen design. The mechanical property parameters of GFRP pipe: E1=43.15GPa, E2=E3=14.32GPa, ν12= ν12=0.283, ν23=0.091, G12=G31=2.48GPa. The mechanical property parameters of steel sleeve: E=201.5GPa, ν=0.241. The mechanical property parameters of adhesive: E=5GPa, ν=0.25. The bond strength between steel and adhesive is 16.5MPa, and the bond strength between GFRP and adhesive is far greater than that between the steel and adhesive, so the failure of the connection between GFRP and adhesive is not considered. The geometric model established is shown in figure 7 and the details of the binding site are shown in figure 8. Because the steel sleeve model is complex, it is divided by tetrahedral mesh. GFRP and adhesive are divided into hexahedral meshes. The finite element model is shown in the figure 9. Because the specimen is short column, the effect of slenderness ratio is not considered. The constraint conditions of compression and tensile simulation are all fixed on one end, and the longitudinal constraint is relaxed and the longitudinal axial force is exerted on the other end.

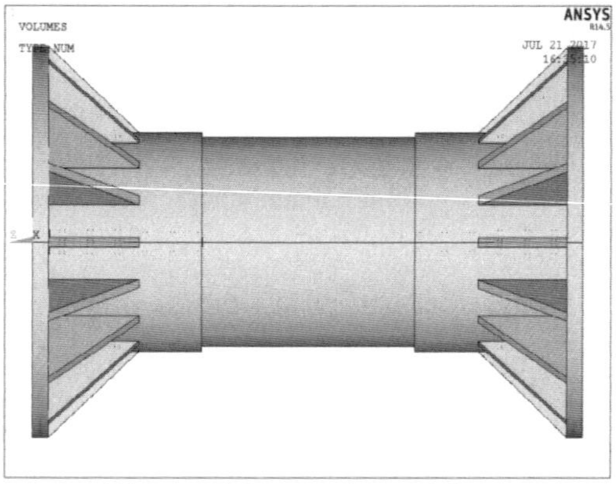

Figure 7 Geometric model

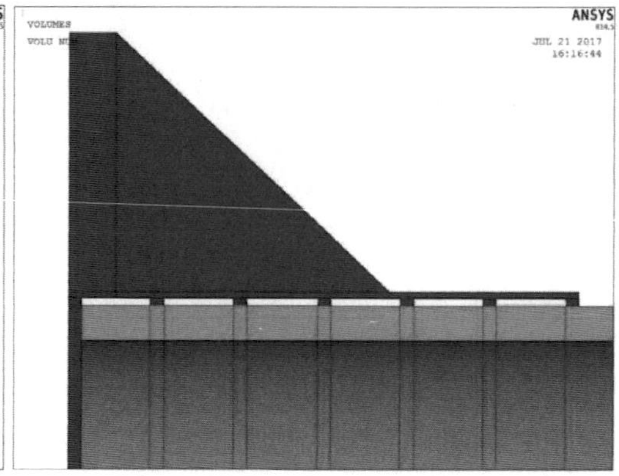

Figure 8 Details of the binding site

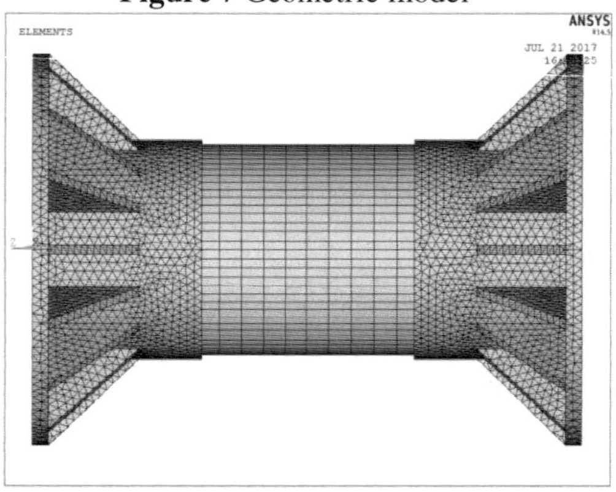

Figure 9 The finite element model

As for the axial compression component, the ultimate bearing capacity of finite element simulation is 6688kN. The comparison of load–displacement curves between experimental and FE result are shown in figure 10. It can be seen from the figure that the finite element analysis results are similar to that of the actual test, which demonstrates the finite element analysis method is valid.

As for the axial tension component, the ultimate bearing capacity of finite element simulation is 4589kN. The load-displacement curve of the finite element analysis is shown in figure 11. The tension ultimate bearing capacity is less than the compression capacity, which is in line with the common sense of gusset connection. On the one hand, it also shows the correctness of simulation.

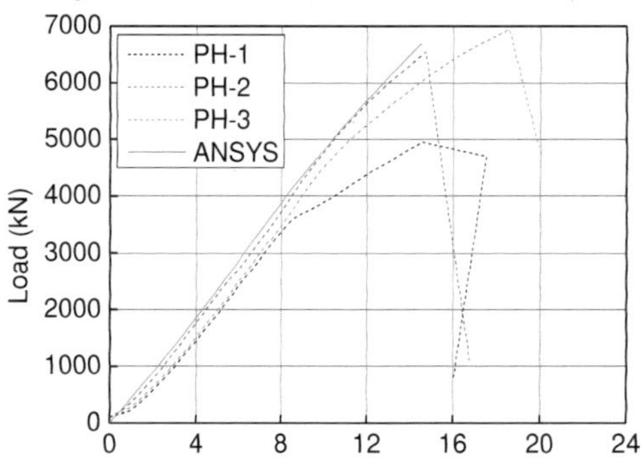

Figure 10 Comparison of load–displacement curves between experimental and FE result

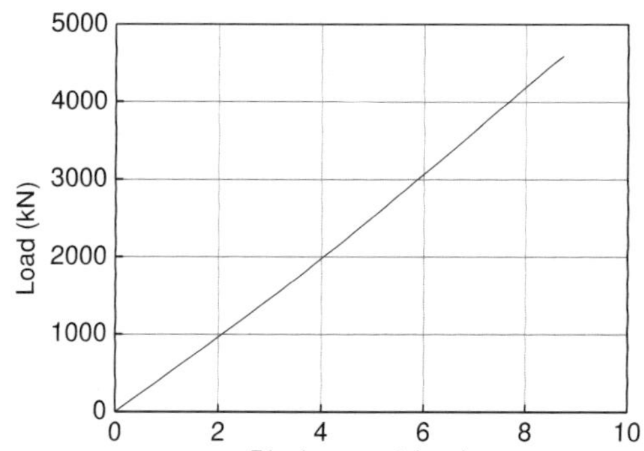

Figure 11 Load-displacement curve of tension FE result

6. Conclusion

In this paper, axial compression experiment of the connection property between full-scale composite member and steel sleeve was conducted. The tensile connection strength of the component was obtained through the FE method. The main conclusions are as follows:

(1) As for the axial compression experiment, the ultimate bearing capacity of the component is between 4950kN and 6950kN, the mean value is 6150kN. The standard deviation of the test data of the ultimate bearing capacity of the component is 1058.3kN, which shows that the dispersion of connection performance is quite large. The longitudinal strain of the steel sleeve connecting the fiberglass composite material is about 8000 to 10000 $\mu\varepsilon$, the steel sleeve strain is less than 800 $\mu\varepsilon$, which is still in the elastic stage.

(2) As for the tension analysis, the ultimate bearing capacity of the component is 4589kN. The failure of the component also results from the failure of the adhesive. The tensile ultimate bearing capacity of the component is less than the extrusion bearing capacity.

(3) Bonded steel sleeve connection for full-scale composite member can better exert the strength of the composite material, for the connection strength is close to the compression strength of GFRP. The connection mode is capable to meet general engineering requirement.

7. References

[1] Bakis C 2002 Fiber-Reinforced Polymer Composites for Construction—State-of-the-Art Review *J Composites for Construction.* 6(2):73-87.

[2] Hollaway L C, 2010 A review of the present and future utilisation of FRP composites in the civil infrastructure with reference to their important in-service properties *J Construction & Building Materials.* 24(12):2419-2445.

[3] Demers C E 1998 Fatigue strength degradation of E-glass FRP composites and carbon FRP composites *J Construction and Building Material.* 12(5): 311-318.

[4] Shi X, Wang C 2011 Application of Glass FRP Anti-corrosion Base Construction Method in Power Transmission and Transformation Projects *J Jilin Electric Power.* 1574-1579 88(11):págs.

[5] Tang B, Wen Y F, Deng J G 2010 Research of a novel type electric power tower made in FRP *J Advanced Technology of Electrical Engineering & Energy.* 80(3):16-215.

[6] Wang H, Jianmin H U, Zhao X 2011 Stability Analysis of Axial Compression Component of FRP *J Electric Power Construction.*

[7] Nagaraj V, Gangarao H V S 1997 Static Behavior of Pultruded GFRP Beams *J Composites for Construction.* 1(3):120-129.

[8] Tommaso A D, Russo S, 2003 Shape Influence in Buckling of GFRP Pultruded Columns *J Mechanics of Composite Materials.* 39(4):329-340.

[9] Keller T, Schollmayer M 2004 Plate bending behavior of a pultruded GFRP bridge deck system *J Composite Structures.* 64(3):285-295.

[10] Qiu C, Feng P, Yang Y 2017 Joint capacity of bonded sleeve connections for tubular fibre reinforced polymer members *J Composite Structures.* 163:267-279.

IOP Conf. Series: Earth and Environmental Science **114** (2018) 012018 doi:10.1088/1755-1315/114/1/012018

Slicer Method Comparison Using Open-source 3D Printer

M K A Mohd Ariffin[1a], N A Sukindar[2], B.T. H T Baharudin[1], C N A Jaafar[1], and M I S Ismail[1]

[1]Department of Mechanical and Manufacturing University Putra Malaysia, Malaysia
[2]Mechanical Department of Politeknik Kuching, Sarawak, Malaysia

[a]khairol@upm.edu.my

Abstract. Open-source 3D printer has been one of the popular choices in fabricating 3D models. This technology is easily accessible and low in cost. However, several studies have been made to improve the performance of this low-cost technology in term of the accuracy of the parts finish. This study is focusing on the selection of slicer mode between CuraEngine and Slic3r. The effect on this slicer has been observe in terms of accuracy and surface visualization. The result shows that if the accuracy is the top priority, CuraEngine is the better option to use as contribute more accuracy as well as less filament is needed compared to the Slice3r. Slice3r may be very useful for complicated parts such as hanging structure due to excessive material which act as support material. The study provides basic platform for the user to have an idea which option to be used in fabricating 3D model.

1. Introduction

The first paragraph after a heading is not indented (Bodytext style). 3D Printing technology has been widely used since the expiration of FDM technology a few years ago [1]. The first open-source 3D printer has been developed by Jones *et al* and the first model was named as Version I *"Darwin"*[2]. Nowadays, the popular open-source technology is including MakerBot and Ultimaker which commonly used use polymer materials such as acrylonitrile butadiene styrene (ABS) and polylactic acid (PLA) [3]. The cost of this technology is around $1500 to $5000 which is quite affordable [3].

Low-cost is one of the factor that this open-source 3D printing has been chosen to fabricate the 3D models. However, the performance of open-source 3D printer still in doubt where study was done on the dimensional accuracy using low-cost 3D printer shows differential in dimension to the actual value [4]. Also other study showing 20% different between the finish parts and the CAD data [5]. Despite of this drawback, some study was done to improve the performance by investigating the process parameters to increase the accuracy of the finish parts [6]. Other study was done on improving the nozzle with specific die angle to improve the accuracy as well [7]. Cost also has been study by comparing several 3D printing technologies to give an idea in choosing the right parameters and technology depending on desired specification [5].

This present study focus on comparing the slicer mode to observe the different on the performance. The slicing option is to convert the design model from CAD data into readable code to start the printing process. So far, there is no study has been done regarding the selection of slicer mode option and their effect on the finish parts. This study will observe the two slicers which are CuraEngine and Slic3r to see the effect on the finish parts which can be used to improve the product.

2. Methodology

A new open-source was developed in the lab consist of micro-controller (Arduino), having five stepper motor for x, y, z axes and for the feeding filament. The 3D printer using lead screw for all the three axes and also single power supply. The build dimension is 21mm x 21mm x 30mm.

2.1 Slicing Method

In this research, two types of slicer have been used for the comparison which are CuraEngine and also Slic3r. The function of this slice is to convert the design model from CAD data into readable code such G-code for the printing process. This process need to be done after setting up the parameters such layer thickness, percentage infill, printing speed and so on. The slicing software will automatically convert all the data into readable code for the extrusion process.

2.2 Sample Fabrication

In order to compare the slicing method, two samples were designed using Autodesk Inventor Software (Autodesk, USA) and converted into standard triangular language (STL) format. The first sample is quite simple with straight and curve design as shown in Figure 1(a) while the second sample more complicated with a combination of roundness shape, chamfer, fillet, and tiny hole as depicted in Figure 1(b).

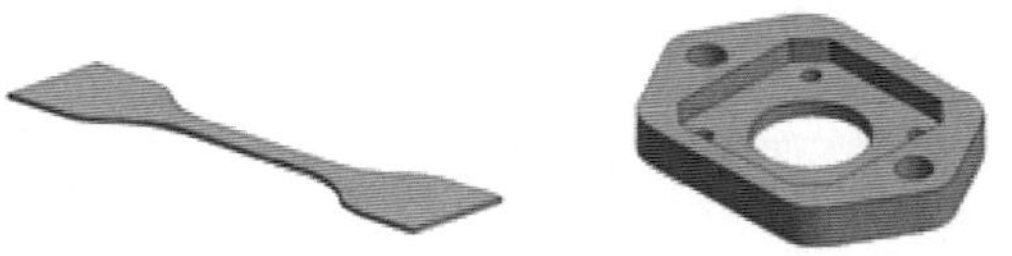

a b

Figure 1. (a) Sample 1 (b) Sample 2

The process parameters were fixed for both slicer including layer thickness of 0.3 mm, shell thickness of 1.2 mm, printing speed of 60 mm/s, and rectilinear fill pattern.

2.3 Dimensional Accuracy Measurements

The dimension was measured only for sample 1 using vernier calliper. Several dimension on the sample parts was taken to compare the differences for both slicer CuraEngine and Slic3r.

3. Results and Discussion

The study focus on two aspects which are the dimensional accuracy and the surface visual of the finish parts. There were two parts from sample 1 were fabricated by using both slicer. This condition was applied to the sample 2.

3.1 Accuracy Measurements

Two parts from sample 1 were printed using CuraEngine and Slic3r slicer as shown in Figure 2. Both slicer was compared and the data were recorded in Table 1.

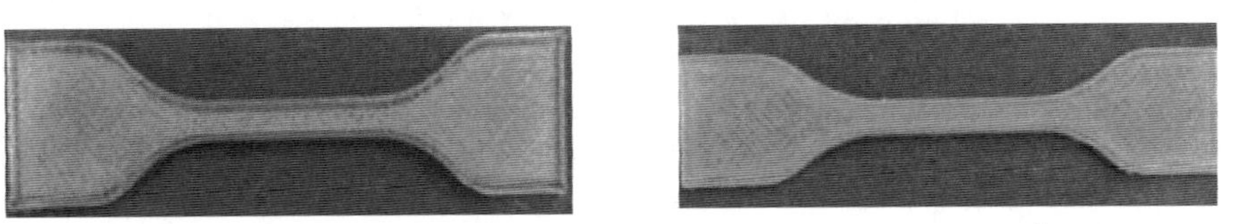

a b

Figure 2. Printed sample 1 (a) Slice with CuraEngine; (b) Slice with Slic3r

Table 1. Dimension measurements for both slicer

Nominal Dimension (mm)	CuraEngine (mm)	Slic3r (mm)
25.0	25.0	24.9
6.2	6.2	6.0
2.0	2.0	1.9
113.5	113.4	113.3

Based on the measurement of the dimension, there is slightly different between slice with CuraEngine and Slic3r. Overall, the CuraEngine give much better accuracy around 0.1 mm. Some of the feature shows quite significant different around 0.2 mm.

3.2 Surface Visual

The sample 2 design was exported to Repetier-host software. The first model was sliced using option of CuraEngine. The second model from the same sample 2 was sliced using Slic3r. To compare both slicer, two models from sample 2 were fabricated. By visualizing the surface of the finished parts, there is a difference in terms of the smoothness of the certain parts. If we observe carefully, especially in roundness shape and the two-tiny hole showing some of the excessive materials by using Slic3r while the CuraEngine just slightly excessive material. This clearly can be seen in Figure 3 below.

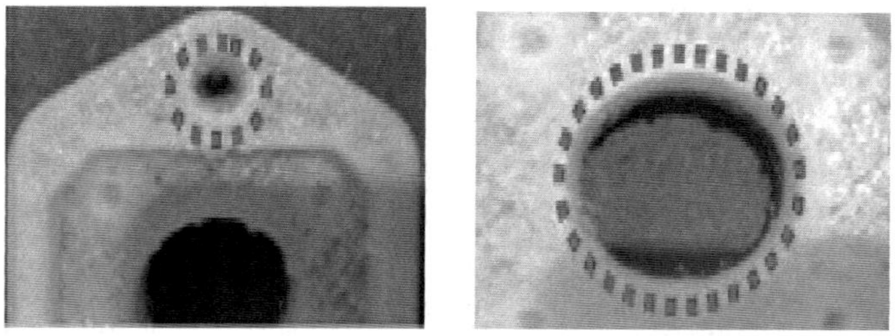

a b

Figure 3. (a) Excessive material in tiny hole (b) Excessive material in roundness shape

For the finished parts that slice using CuraEngine, the model smoother compare to the Slic3r and just a slight excessive material as shown in Figure 4.

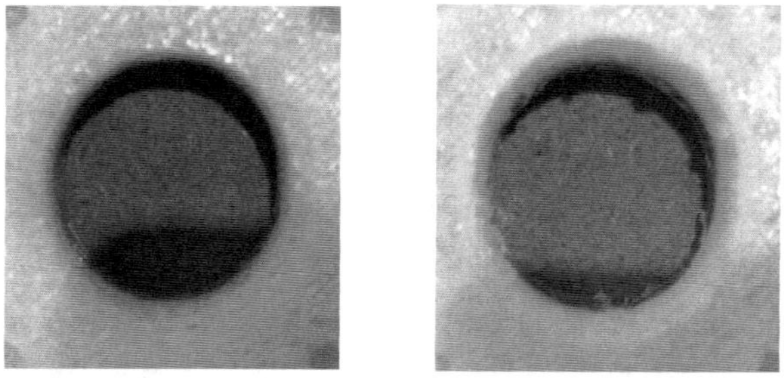

a b

Figure 4. (a) Smooth parts slice using CuraEngine (b) Excessive material appear using Slic3r

Table 2. Comparison between CuraEngine and Slic3r

Printing Spec	CuraEngine (mm)	Slic3r (mm)
Printing time	38m 19s	35m 23s
Layer Count	29	33
Filament needed	6230 mm	6968 mm

The finding from this study is that if the models need more accuracy, the better slicer option was to use the CuraEngine. In addition less time required and save more filament which is less cost as shown in Table 2. Besides, the finished parts are also good enough which is within the range of 0.1 mm. Meanwhile, if using Slic3r, more filament was needed compared to the CuraEngine which shows a lot of excessive materials appeared on the finish parts. In terms of accuracy, the Slic3r less accurate due to excessive material. However, Slice3r maybe is the best option for the complicated parts which involve a lot hanging structure because the excessive material can be used as support material.

4. Conclusion

Both slicer between CuraEngine and Slic3r has been compared in terms of dimensional accuracy and the surface visualization. It shows that CuraEngine contribute more accuracy and less filament is needed for printing the parts. However, the Slice3r can be used for parts that possess a lot of hanging structure. Overall, the study is providing the platform for the user to have many options available so that they can manipulate to have any output that they wanted depending on the spec.

5. Acknowledgment

The research was funded by the Fundamental Research Grant Scheme under the Ministry of Higher Education Malaysia (MOHE) FRGS/1/2015/TK03/UPM/02/3 under vote number 5524728.

References

[1] Crump, S. S., "Apparatus and method for creating three-dimensional objects." U.S. Patent 5,121,329, issued June 9, 1992.

[2] Jones, R., Haufe, P., Sells, E., Iravani, P., Olliver, V., Palmer, C., and Bowyer, A. 2011. RepRap – the replicating rapid prototyper. Robotica. 29: 177–191.

[3] B. N. Turner, R. Strong, and S. A. Gold, "A review of melt extrusion additive manufacturing processes: I. Process design and modeling," *Rapid Prototyp. J.*, vol. 20, no. 3, pp. 192–204, 2014.

[4] G. W. Melenka, J. S. Schofield, M. R. Dawson, and J. P. Carey, "Evaluation of dimensional accuracy and material properties of the MakerBot 3D desktop printer," *Rapid Prototyp. J.*, vol. 21, no. 5, pp. 618–627, 2015.

[5] D. a. Roberson, D. Espalin, and R. B. Wicker, "3D printer selection: A decision-making evaluation and ranking model," *Virtual Phys. Prototyp.*, vol. 8, no. 3, pp. 201–212, 2013.

[6] L. M. Galantucci, I. Bodi, J. Kacani, and F. Lavecchia, "Analysis of Dimensional Performance for a 3D Open-source Printer Based on Fused Deposition Modeling Technique," *Procedia CIRP*, vol. 28, pp. 82–87, 2015.

[7] N. A. Sukindar, M. K. A. Mohd Ariffin, B.T. H. T. Baharudin, C. N. A. Jaafar, M. I. S. Ismail, "Comparison on Dimensional Accuracy Using a Newly Developed Nozzle for Open-Source 3D Printer," *Appl. Mech. Mater.*, vol. 859, pp. 15–19, 2016.

IOP Conf. Series: Earth and Environmental Science **114** (2018) 012019 doi:10.1088/1755-1315/114/1/012019

Design of external vibration absorber for vibration suppression of milling cutter in processing

ZHANG Ya-hui[1], ZHANG Nian-song[1] and WANG Ai-min[2]

[1] School of Mechanical Engineering, Nanjing University of Science and Technology, Nanjing 210000, China
[2] School of Mechanical Engineering, Beijing Institute of Technology, Beijing 100081, China

18018028360@163.com

Abstract. A new type of external dynamic vibration absorber is designed to control the vibration at a specific frequency of the milling cutter during the milling process. The structural design of the dynamic vibration absorber and the selection of the corresponding parameters are conducted. The finite element model of the cutter is established and connected with the vibration absorber. The results of the harmonic response analysis of milling cutter before and after the installation of the vibration absorber are compared and show that the vibration absorber can reduce the vibration of the cutter at the resonant frequency, which means it has a good vibration damping performance. The vibration absorber has the advantages of simple structure, convenient frequency modulation and easy installation. This context lay the foundation of further application for damping cutter.

1. Introduction

Because of the high efficiency of processing, high-speed milling has accounted for more and more in the machinery manufacturing industry in recent years. Process the aerospace heterogeneous thin-walled structural parts usually requires milling cutter with a long tool bar. Because of the large aspect ratio of the milling cutter, it is very easy to produce vibration during processing, which will reduce the machining precision of the parts and will restrict the processing efficiency seriously.

The dynamic vibration absorber is used by adding the vibration absorber substructure to the main system to change the vibration response of the main system. The development and application of the vibration absorber has a history of over one hundred years. In the aspect of control mode, it is mainly divided into passive, semi-active and active dynamic vibration absorbers [1, 2], in the aspect of freedom, it develops from single degree of freedom to multi degrees of freedom [3, 4]. Dynamic vibration absorber has been applied in the control of vibration, Handong used built-in dynamic vibration absorber to reduce the cutter vibration of the large aspect ratio cutter, and obtain the desired effect through the experiment [5]. Yang et al. Proposed a long-span milling cutter structure with a built-in two-degree-of-freedom damper, experiments show that the surface quality of the parts can be significantly improved [6]; Zhang Hengming et al. studied the vibration reduction effect of various kinds of milling cutters filled with different parameters of damping particles [7]; Madoliat improved the damping performance of the cutter by applying friction damping to long end mill [8]. Wang et al [9] designed a nonlinear single-degree-of-freedom dynamic vibration absorber, modal test and cutting test showed good vibration suppression effect. The silent vibration reduction tool series of Sweden Sandvik Company is dominant in the market, and its aspect ratio can reach more than 12 times [10].

In this paper, based on the dynamic vibration absorption method, the dynamic vibration absorber is used to reduce the vibration of the cutter, and it is placed on the milling cutter bar externally. A new vibration reduction scheme of milling cutter is designed for a certain type of milling cutter bar. The finite element simulation were carried out to verify the effectiveness of the vibration absorber.

2. Damping principle of dynamic vibration absorber for milling cutter

The dynamic vibration absorber is a vibration system composed of mass element, spring element and damping element, usually mounted directly on the target vibration system, its working mechanism is the vibration of the main structure will lead to the vibration of absorber, the reacting force generated by the absorber will transmit to the main vibration system to control the vibration of the main vibration system to a certain extent. The additional absorber is called a subsystem, which forms a two-degree-of-freedom system with the main vibration system. The simplified model is shown in figure 1.

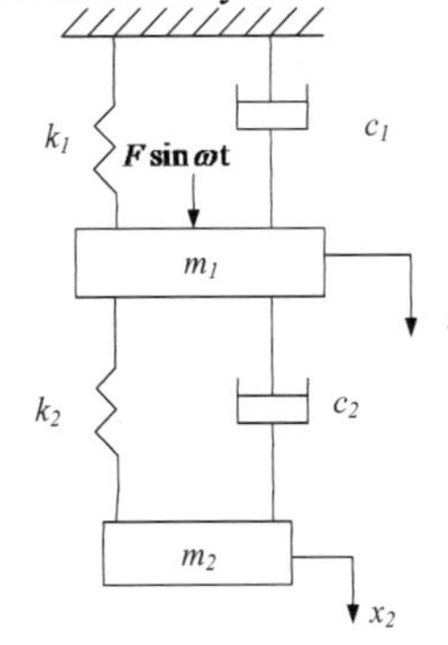

Figure 1. Two degrees of freedom system model.

Where F is the external load; m1, k1, c1 are the mass, stiffness and damping of the main system respectively; m2, k2, c2 are the mass, spring stiffness and damping of the dynamic vibration absorber respectively; x1, x2 represent the displacement of main system and subsystem, the kinematics equation is shown as equation (1).

$$\begin{cases} m_1\ddot{x_1} + (c_1 + c_2)\dot{x_1} - c_2\dot{x_2} + (k_1 + k_2)x_1 - k_2x_2 = F \\ m_2\ddot{x_2} - c_2\dot{x_1} + c_2\dot{x_2} - k_2x_1 + k_2x_2 = 0 \end{cases} \quad (1)$$

The fixed-point theory was developed in the last century when the optimal design of the dynamic vibration absorber was studied. It is a method of designing a vibration-damping device using a specific point on the frequency response function curve which is independent of the damping.

The optimal design formula of the dynamic vibration absorber is obtained after a long period of development and promotion [11]. Among them, the dynamic vibration absorber and the main system of the natural frequency ratio need to meet the optimal homology condition:

$$\gamma = \omega_2/\omega_1 = 1/(1+\mu) \quad (2)$$

Damping ratio needs to satisfy optimal damping conditions

$$\zeta = \frac{3\mu}{8\ (1+\mu)^3}^{1/2} \quad (3)$$

Where μ is the mass ratio of the dynamic vibration absorber to the main system.

3. Structure and parameter determination of vibration absorber based on finite element simulation

3.1. Modal analysis of milling cutter based on finite element simulation

Milling cutter is a multi-degree of freedom system, which has a multi-order natural frequency and the corresponding vibration mode in theory. When the adjacent natural frequency is far apart, a single natural frequency and its vibration mode can be chosen as the target. Under the actual working conditions, the working frequency is closest to the first-order frequency of the milling cutter, then the first-order mode of the milling cutter is taken as the research object, which is simplified as a single-degree-of-freedom system model. The finite element simulation software can easily and efficiently obtain system model's natural frequency and its corresponding vibration mode by modal analysis. In this paper, abaqus software is used to analyze and the DHX's X-WDEX-200150 type milling cutter is used, which is mainly composed of a cutter head and a cutter bar. The cutter bar's diameter is 20mm, length is 150mm, the quality is 0.743kg, the material is tungsten steel which has a good comprehensive mechanical performance. The model is shown in figure 2.

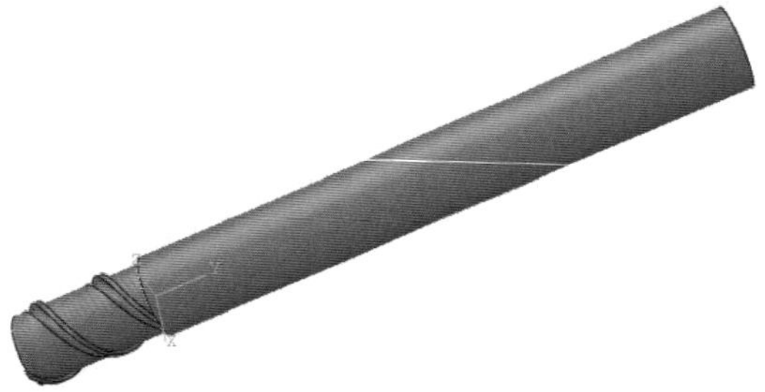

Figure 2. Three-dimensional model of milling cutter.

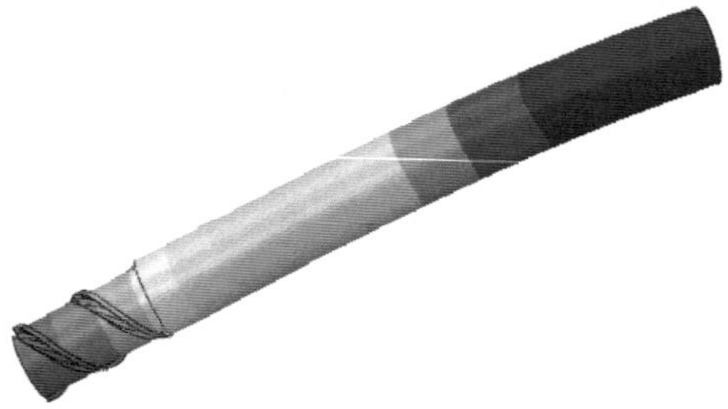

Figure 3. First order vibration pattern of milling cutter.

The model of the milling cutter is imported into the simulation software abaqus, and the material parameters, connection conditions and boundary conditions are set, then mesh and do modal analysis. The result shows that 592.88hz is the only natural frequency below 1000hz, the corresponding mode is shown in figure 3.As can be seen from figure 3, the first order vibration mode of the milling cutter is mainly the bending deformation of the milling cutter along the radial direction.

3.2. Structural design of dynamic vibration absorber

As the cutter arbor is cylindrical, the radial vibration may be in any direction, and the dynamic vibration absorber generally has only one direction of freedom, that is, only one direction has control effect. So in this article, the vibration absorber for the cutter is designed as a circular dynamic vibration absorber with vibration absorb performance in the circumferential direction. The structure is shown in figure 4, the vibration absorber is designed as a mass block fixed on a cantilever beam, and the direction of the cantilever beam is arranged along the axial direction of the cutter bar, so that the installation space can be greatly reduced. The vibration absorber is connected with the cutter arbor by two semi-circle parts. By adjusting the position of the mass block on the cantilever beam, the natural frequency of the vibration absorber can be changed.

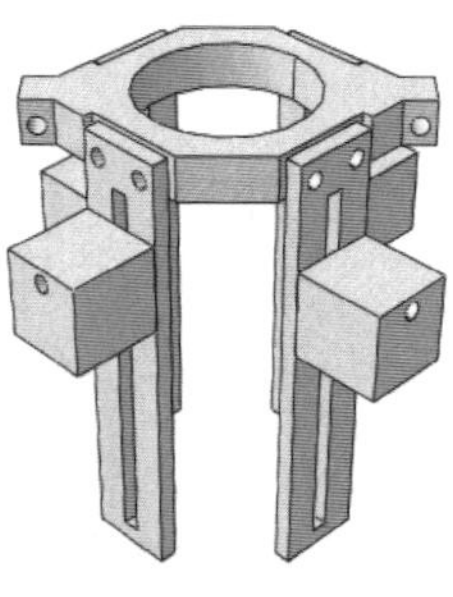

Figure 4. Model of vibration absorber.

As the vibration absorber is installed outside the milling cutter, the installation space is limited, and it is necessary to reduce the size of the vibration absorber as much as possible. So the mass block is chosen as W-Cu alloy, the density is 14000 kg / m3, the elastic modulus E = 331200mpa, Poisson's ratio is 0.318; the material of cantilever beam is selected as spring steel, the density is 7800kg / m3, elastic modulus E = 197000mpa, Poisson's ratio is 0.23. The material of the part which connects the vibration absorber and tool is 6061 alloy aluminum, the density is 2800kg / m3, the elastic modulus E = 68900mpa, Poisson's ratio is 0.33.

3.3. Parameter determination of vibration absorber based on finite element simulation

Obtain the original milling cutter system quality is 0.743kg, the target frequency is 592.88hz, the stiffness is 10310557N / m through the simulation. The mass of the absorber is chosen as 0.04kg, then the mass of each single block is 0.01kg, and the optimal stiffness of each single absorber is obtained according to equation (4).

$$k_i = \mu_i {\gamma_i}^2 k_0 \quad \mathrm{i} = 1,2,3,4 \tag{4}$$

Thus, the parameters of the vibration absorber designed in this paper can be obtained as shown in table 1.

Table 1. Parameters of vibration absorbers.

	Mass[kg]	Stiffness[N/m]	Frequency[hz]
Single absorber 1	0.01	99520	502.08
Single absorber 2	0.01	136750	588.55
Single absorber 3	0.01	100030	503.37
Single absorber 4	0.01	137210	589.53

3.4. Selection of structural parameters for dynamic vibration absorbers

A single vibration absorber is simplified as a vibration system composed of a centralized mass block and cantilever for the purpose of frequency analysis. The inherent frequency of the system which equipped with mass block and neglect the mass of cantilever is

$$\mathrm{f} = \frac{1}{2\pi}\left(\frac{3EI}{l^3 m}\right)^{1/2} \tag{5}$$

The natural frequency of the cantilever beam system without mass block is

$$\mathrm{f} = \frac{0.356}{2\pi}\left(\frac{\pi}{l}\right)^2\left(\frac{EI}{\rho_1}\right)^{1/2} \tag{6}$$

When the two systems are combined, the vibration frequency is

$$\mathrm{f} = \frac{1}{2\pi}\left(\frac{3EI}{l^3\ (m+M)}\right)^{1/2} \tag{7}$$

When the mass block slides on the cantilever beam, the first-order natural frequency of the single-absorber system is

$$\mathrm{f} = \frac{1}{2\pi}\left(\frac{3EI}{(tl)^3(m+\frac{M}{t^3})}\right)^{1/2} \tag{8}$$

Where $t = l_0/l, M = 3l\rho_1/0.356^2\pi^4$, $I = ab^3/12$, l_0 is the length between the centroid of the mass block and the fixed end of the cantilever beam; l is the length of the cantilever beam; E is the elastic modulus of the cantilever beam material; I is the inertia moment of the cantilever beam; a and b are the width and height of the cantilever beam's cross section respectively; m is the mass of the mass block, M is the equivalent mass of the beam at end position, and ρ_1 is the linear density of the cantilever beam.

According to equation (4), equation (8), the dimensions of the components designed in this paper are shown in table 2.

Table 2. Dimensions of each component of the vibration absorber.

mass block	length*width*height[mm*mm*mm]	10*10*10
cantilever beam	length*width*height[mm*mm*mm]	50*10*2
fixed clamping component	inner diameter[mm]	20

The relationship between the change of the parameters and the frequency of vibration absorber is calculated by MATLAB. The relation between the position of mass block on the cantilever beam and the frequency is shown in figure 5.

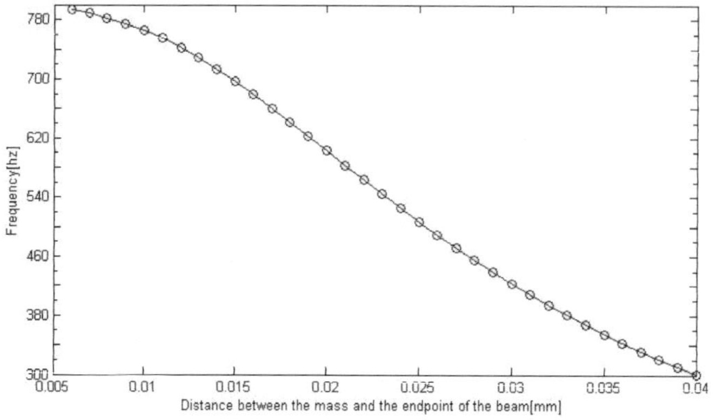

Figure 5. The relationship between the length of cantilever beam and the frequency of absorber.

4. Simulation of vibration reduction effect based on finite element simulation

4.1. Modal analysis of dynamic vibration absorber

In According to the dimensions of the components of the absorber obtained by Eq. 8, the position of the mass block on cantilever beam is adjusted in the solidworks software to obtain the desired vibration absorber frequency. Then, the adjusted vibration absorber model is imported into abaqus for modal analysis, and the first four modes are shown in figure 6.

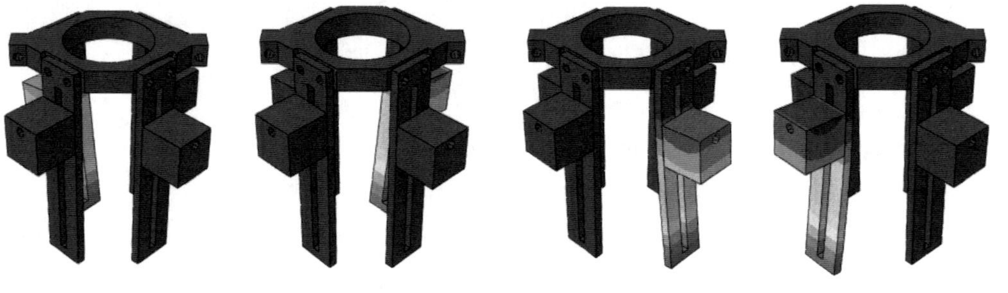

Figure 6. First four modes of vibration absorber.

4.2. Modeling of vibration reduction milling cutter with additional dynamic vibration absorber

Connect the vibration absorber with the milling cutter and import into the ABAQUS, as shown in figure 7. The whole assembly is pre-processed and submitted for harmonic response analysis. The amplitude and frequency curves of the response point before and after the installation of the absorber are shown in figure 8.

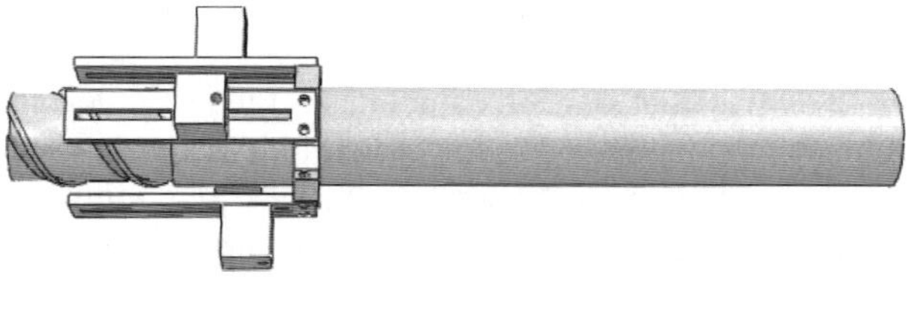

Figure 7. Overall drawing of milling cutter.

IOP Conf. Series: Earth and Environmental Science **114** (2018) 012019 doi:10.1088/1755-1315/114/1/012019

4.3. Analysis of vibration reduction effect
As can be seen from figure 8, the amplitude of the vibration at the response point after installing the vibration absorber is reduced by 48.2% compared to the condition without the absorber. At the same time, it can be seen from the overall simulation diagram showed in figure 9, the vibration amplitude on the milling cutter becomes smaller and the vibration is transmitted to the vibration absorber.

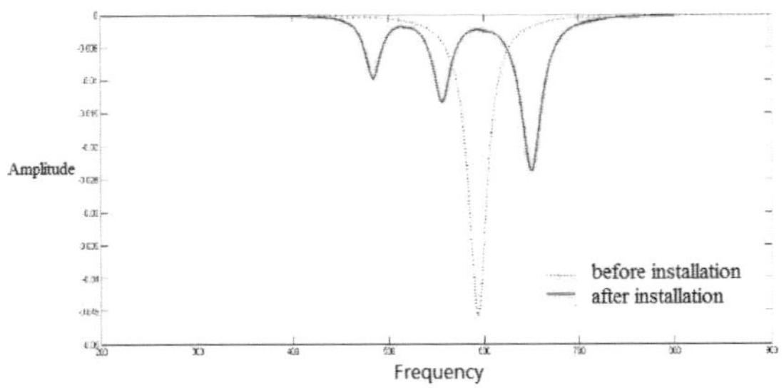

Figure 8. Amplitude frequency curve of response point before and after the installation of vibration absorber.

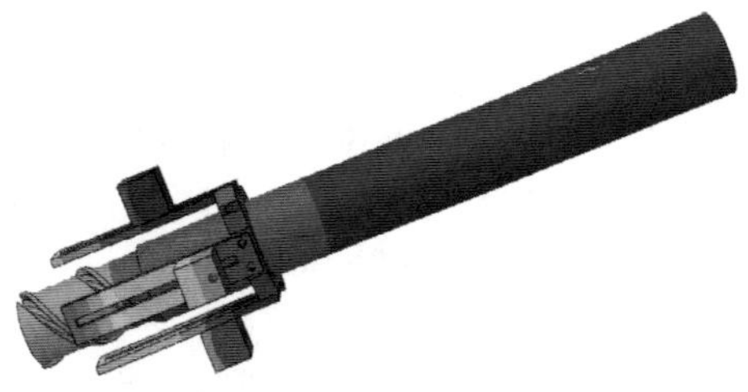

Figure 9. Frequency response of milling cutter after installing dynamic vibration absorber.

5. Conclusion
In the milling process, the vibration of the tool often causes a series of adverse effects, the existing solution is to hollow the milling cutter arbor and equip with built-in absorbers. This method is difficult to realize and can cause damage to the cutter arbor. Aiming at the above problems, this paper designed an external vibration absorber and apply it to the cutter vibration control, the structure of the cantilever beam has the advantages of simple structure, convenient installation and easy adjustment. The simulation shows that the vibration can be reduced by 48.2% at the target frequency, the dynamic vibration absorber has good vibration reduction effect, and lays a foundation for the application of the damping milling cutter.

References
[1] Liu Yaozong, Dianlong Yu, Zhao Honggang and Wen XISEN 2007 *Chin J. Mech eng.* **03** 14
[2] Chen Zili, Ma Aoling, Yu Gang and Chen Changping 2012 *J. Lujiang Univ.* **03** 67
[3] Tian Hongliang, Zhou Yang, Zheng Jinhua and Chen Tianmin 2015 *J. Ctgu.* **06** 91
[4] Li Jun and Liu Jianhua 2003 *J.Jovmd.* **02** 23
[5] Han Dong 2017 *HUST*
[6] Yang Yi-qing and Yu Yu 2016 *J.COMPUT INTEGRA MANUF SYS.* **11** 2588
[7] Zhang Heng-ming and Pang Xue-hui 2016 *J. MT&MT.* **06** 89
[8] Madoliat R, Hayati S and Ghalebahman A G 2011 *J.Jamdsm.* **2** 115
[9] Wang M, Zan T and Yang Y 2010 *J.Ijmtm.* **5** 474
[10] Seco Tools Jabro JC800 Series Award 2012 *J.Tools Technology.* **02** 4
[11] Kazuto Seto 2013 *Dynamic vibration absorber and its applications* (Beijing: Beijing Machinery industry press)

IOP Conf. Series: Earth and Environmental Science **114** (2018) 012020 doi:10.1088/1755-1315/114/1/012020

Designing Universitas Indonesia Molina EV Bus Dashboard Using ECQFD and TRIZ

Muhammad Faiq Pradhila[1,1], Amalia Suzianti[1,2] and Prilly Putri Adinda[1,3]

[1]Industrial Engineering Department, Faculty of Engineering, Universitas Indonesia, Depok, Indonesia

[1]mhd.faiq@ui.ac.id, [2]suzianti@eng.ui.ac.id and [3]prilly.putri@ui.ac.id

Abstract. Universitas Indonesia is involved in the national electric car development program. One of the focus by the research team is to develop the Molina EV Bus which is planned to replace the current operational bus at UI so that it can be more environmental friendly. With UI developing facilities for the disabled, the Molina research team planned to make a new prototype of the Molina EV Bus to contribute to the facilities developed for the disabled. The new prototype is expected to increase the quality of the previous features of the EV Bus, including the dashboard that had been ignored. To support the development of the new prototype, this research was conducted to design a suitable dashboard for the new prototype. Design of the prototype are made using Autodesk Inventor. This research used the integration of ECQFD (Environmentally Conscious Quality Function Deployment) and TRIZ (Theory of Inventive Problem Solving) method. ECQFD was used to translate user needs into quality characteristics based on environmental aspects. TRIZ was used to translate the quality characteristics into technical specifications. This research has generated 3 sustainable, innovative, and user-preferred dashboard design recommendation for the new prototype.

1. Introduction

It is inevitable that today, most of the people in Indonesia love to buy a motor vehicle, whether it is a motorcycle, car, or other. This is one of the contributing factors of the increasing number of motor vehicles in Indonesia each year [1]. According to data, about 32,7% of energy consumption patterns in Indonesia in 2011 came from fuel oil. The increase pattern of fuel consumption in Indonesia every year is one of the consequences of the habits of people who love to buy motor vehicles. In the event of a scarcity of fuel, this large dependence on fuel will cause an energy crisis.

With the aim to reduce emission, the automotive industry has started to divert the source of power generated for car engines by conducting research and development [2]. This new source of power is called electric motor, which do not generate emissions at all. Cars with the electric-based engines are cares that are expected to play an important role to help keep the environment in the future. Indonesia is also active to develop electric cars. Research and development has also been done both by ministries and also by universities. A total of five State Universities (PTN) are appointed to develop a research roadmap, with Universitas Indonesia (UI) being one of them. The mandate of the development of electric cars was marked by the signing of a cooperation mandate between the five state universities, with the Institute of Education Fund Management (LPDP).

Research at Universitas Indonesia on electric cars is still running. Universitas Indonesia (UI) has launched 4 units of electric vehicles. Compared with the other four universities that focus on the technical development of small urban vehicles (city car), UI also examines the technology of electric buses and social studies related to the acceptance of society and industry in the homeland of electric vehicles in Indonesia.

When viewed directly on the prototype of this bus, the dashboard has not received attention and is still using the standard design of the car body maker. Meanwhile, this bus is a convertible electric bus which means that the device from this bus dashboard could be different from the required functionality required on an electric car.

Currently, planning is underway for the development of the 2nd Molina EV Bus prototype. The prototype was developed with the aim of updating the technology to be used on this bus. In addition, the prototype was developed to improve facilities and infrastructure at Universitas Indonesia, specifically in the field of infrastructure for disabled users. Therefore, a new bus design to support those needs is required. The different prototypes of the new UI electric bus are planned to have low-floor features. In addition, it is also expected for the dashboard part of the bus to be updated.

2. Theoretical Overview

2.1. Electric Car
An electric car is a car that is driven by an electric motor by utilizing electrical energy stored on rechargeable batteries or other energy storage media as a source of energy. The use of electric motors provides instant torque to electric cars resulting in a smooth and strong acceleration.

Along with its development, electric car technology is also applied to more large-capacity cars such as buses. Just like an electric car, electric buses are generally divided into hybrid and pure electric. The difference is that electric-powered hybrid buses are still combined with the use of internal combustion engines, while on pure electric buses only use electric motors without conventional combustion engines.

2.2. Dashboard
The car dashboard is the front area of its interior that gets most attention at all times from the driver and the place where the car operational control activities are centered. Dashboard also serves to place the radio/tape/CD, bottle holders, drawers, button control air conditioners, lights and indicators of the vehicle. In addition to function as a panel of various switches and indicators, the dashboard is also a physical boundary between passengers with views beyond the windshield.

The dashboard on the bus has the same function as cars in general, but the dashboard on the bus has more features that do not exist in private cars. This is because the bus has more operational functions compared to private cars.

2.3. QFD (Quality Function Deployment)
QFD is a method for turning qualitative user demand into a quantitative parameter for implementing quality-building functions, and methods for implementing design quality into subsystems, components, as well as specific elements of the manufacturing process [3-8].

2.4. TRIZ
TRIZ (Teoriya Resheniya Izobretatelskikh Zadatach) is an inventive problem-solving method based on logic and data that can help us solve problems creatively. TRIZ is very closely related to contradiction. In product design, contradiction appears when we apply improvement towards a parameter of a system, but at the same time, the other parameter undergoes diminishment. In TRIZ, the contradiction itself is divided into 2 namely contracts of physical and technical contradictions. In general, TRIZ solves technical contradictions, in which occurs between two or more subsystems in a system [8-10].

3. Research Methodology
This section explains about the data and methodology used in this study. This study used two types of data, namely primary data, which was gained through questionnaires, and secondary data, such as data gained from reports, journal, etc. The method used in this study are ECQFD and TRIZ.

In the beginning of data collection, identification of user needs and design attribute is needed [1]. These types of data were obtained using literature study and benchmarking from other buses dashboard, and used to conduct a questionnaire and data processing using ECQFD. Afterwards, questionnaire was obtained based on the user needs and design attributes. This questionnaire aimed to determine the weight of each design attribute based on user preferences [9]. The result of this questionnaire is used as an input data for data processing using ECQFD. In ECQFD, the identification of the relationship between user needs and quality characteristics and between quality characteristics and each component of the product was completed. Table 1. shows the design options obtained to be evaluated.

Table 1. Design Options Comparison

Option 1	Option 2
Component Type	Dimension
Indicator	Structure & Geometry
Component Amount	Weight

Afterwards, QC Improvement Rate, UR Improvement Rate and UR Improvement Effect of each option was calculated. The effect of improvement from each quality characteristic was accumulated to compare between the two options. It was aimed to find which option gives better impact [8].

The result from the comparison showed which option is better than the other. However, that option is not necessarily the best combination to apply improvements to the design. Therefore, in TRIZ data processing, it was needed to find the combination from two design options from ECQFD with the aim to find the best design combination [9].

4. Results and Discussions

This section explains about interpreting the results explained in this study. From the solution of the problem of contradiction with analysis using TRIZ method, solutions for innovation and improvement of bus dashboard design were obtained, namely adjustable body, natural/synthetic fiber composite materials, modular functions, touch screen function, compact shape and separate support devices.

As the result, three design options were obtained to be made for recommendation. In the first design as shown in Figure 1., the main differentiator with other designs is its adjustable function. The dashboard height and proximity can be adjusted with the driver. In addition, the shape and size of this dashboard is compact, thereby it can save space. In this dashboard design, supporting devices such as multimedia and others are located separately from the dashboard.

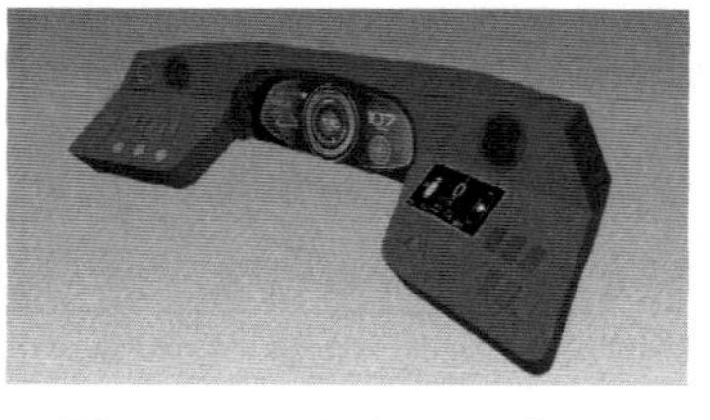

Figure 1. Design Option 1

In the second design shown in Figure 1., the dashboard can function modularly. The desired functions can be easily replaced and installed according to user preferences. There are also supporting devices such as cup holder and multimedia devices in this design.

Figure 2. Design Option 2 *Exploded View*

Fig 2.a. Design Option 2

In the third design, a function that distinguishes this design from other design is that all functions is operated by touch screen. All functions and support devices are also available in touch screen on both side of the dashboard.

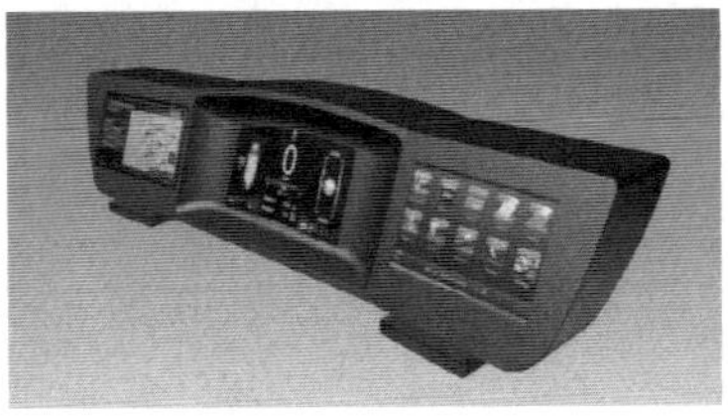

Figure 3. Design Option 3

5. Conclusion

This study aims to generate design factors of consideration for electric bus dashboard design based on user needs, and to recommend dashboard design for the development of the next Molina EV Bus prototype.

The early stage of this study was determining user needs for electric bus dashboard using literature study and interview. Questionnaire was conducted, and the result was used as an input data for ECQFD method. Afterwards, interview and brainstorm with experts were also conducted to process data using ECQFD and TRIZ method.

This study resulted in 6 quality characteristics that can be improved to get an innovative and environmentally friendly dashboard design, namely component type, indicator, component number, dimension, structure and geometry, and weight. Based on troubleshooting contradictions using TRIZ, there are several innovative design specifications, namely adjustable body, natural / synthetic fiber composite materials, modular functions, touch screen function, compact shape, separate support devices and the use of the key as a functioning operating device. This study also resulted in 3 designs of dashboard.

For further research, there are several things to be considered. First, further research can be conducted as deeper design evaluation using ECQFD with involving more environmental factors. Second, the range of this research object can be expanded to public bus, not only UI Molina EV Bus. Third, for a similar study, the methods performed can be integrated with other methods, namely AHP for analyzing the best design option to be chosen, or other ergonomic analytics method to evaluate human factors of the design. Fourth, for further research, survey can be conducted by using another method such as conjoint analysis to find the combination of user-preferred features. In addition, financial and economic factors can be included as concerns for the design to provide an optimal design overview for manufacturing process and consumer purchasing power.

References

[1] Padagannavar P 2016 Automotive Product Design and Development of Car Dashboard Using Quality Function Deployment. *Advances in Automobile Engineering.*

[2] Rathod G, Vinodh S. Madhyasta U Integration of ECQFD and LCA for Enabling Sustainable Product Design in an Electric Vehicle Manufacturing Organization. *Int. J. Sust. Engineering.*

[3] Sakao T 2007 A QFD-centred design methodology for environmentally conscious product product design. *Int. J. of Production Research*

[4] Cohen L 1995 *Quality function deployment: How to make QFD work for you* (New Jersey: Prentice Hall)

[5] Vinodh S and Rathod G 2010 Integration of ECQFD and LCA for Sustainable Product Design *J. Clnr. Prod.* **18** p833-842

[6] Wang J, Yu S, Bin Q and Rao J ECQFD & LCA Based Methodology for Sustainable Product Design.

[7] Wang, J Yu, S Qi, B Rao and J 2010 ECQFD & LCA Based Methodology for Sustainable Product Design. IEEE.

[8] Vinodh S, Kamala V, Jayakrishna K 2013 Integration of ECQFD, TRIZ, and AHP for Innovative and Sustainable Product Development. *Appl. Math. Modell.*

[9] Yudha Irham 2014 Perancangan Desain Laptop Berdasarkan Preferensi Konsumen Dengan Menggunakan Metode ECQFD dan TRIZ (Depok: Program Studi Teknik Industri Fakultas Teknik Universitas Indonesia)

[10] G Altshuller 1999. *The Innovation Algorithm: TRIZ, Systematic Innovation and Technical Creativity* (Worcester, MA: Technical Innovation Centre)